THE NEW MIND READERS

THE
NEW MIND
READERS

*What Neuroimaging Can and Cannot
Reveal about Our Thoughts*

RUSSELL A. POLDRACK

PRINCETON UNIVERSITY PRESS
Princeton and Oxford

Published by Princeton University Press
41 William Street, Princeton, New Jersey 08540
6 Oxford Street, Woodstock, Oxfordshire OX20 1TR

press.princeton.edu

Library of Congress Control Number: 2018937065
ISBN 978-0-691-17861-5

British Library Cataloging-in-Publication Data is available

Editorial: Alison Kalett and Lauren Bucca
Production Editorial: Jenny Wolkowicki
Jacket design: Jason Alejandro
Production: Jacqueline Poirier
Publicity: Sara Henning-Stout
Copyeditor: Maia Vaswani

This book has been composed in ITC New Baskerville

Printed on acid-free paper. ∞

Typeset by Nova Techset Pvt Ltd, Bangalore, India

Printed in the United States of America

10 9 8 7 6 5 4 3 2 1

To my parents, for always supporting my scientific dreams.
And to Mike, who left us far too soon.

CONTENTS

ILLUSTRATIONS

FIGURES

COLOR PLATES (*following page 112*)

ACKNOWLEDGMENTS

I'd like to thank several colleagues whose work and discussion have been particularly helpful in developing this book. Marcus Raichle endured my questions as I wrote the book, and his detailed historical documentation of the early days of neuroimaging in his Society for Neuroscience biography was a gold mine of details about the birth of positron emission tomography (PET) imaging. Steve Petersen also provided valuable comments regarding my treatment of the early days at Washington University. Peter Bandettini read an early draft and gave very helpful comments; in addition, the special issue of *NeuroImage* that he edited in 2012 ("20 Years of fMRI") was uniquely useful in helping to reconstruct the history of the early days of functional magnetic resonance imaging (fMRI). Ken Kwong and Kâmil Uğurbil both provided very helpful feedback on my discussion of the history of the early development of fMRI, and Kâmil also gave useful comments on high-field MRI. Luiz Pessoa provided some very helpful criticism on an early draft, pointing out some ways in which I hadn't actually listened to my own message fully enough, and suggesting a very useful way of thinking about the "dictionary" metaphor for brain decoding. His papers were also the inspiration for figure 1.5. Alex Shackman provided a wealth of comments and helped me more accurately describe the brain's fear systems, as well as pointing me to a number of very useful references. Pietro Pietrini was very helpful with translation of some sections from Angelo Mosso's book in the original Italian. David Kennedy gave me some helpful insights from his birds-eye view of the birth of fMRI at Massachusetts General Hospital in the early 1990s. Joe Devlin and Ken Norman both provided very useful reviews of the entire book.

Thanks are also due to a number of people who supplied helpful discussion and comments on early drafts of the book: Felipe de Brigard, John Bruer, Nico Delaeter, Susan Fitzpatrick, Marta Garrido, David Glahn, Kalanit Grill-Spector, Keith Humphreys, David Kennedy, Anna Khazenzon, Leah Krubitzer, Jamie Li, Dan Lloyd, Tor Wager, and Jochen Weber.

My editor at Princeton University Press, Alison Kalett, helped convince me to write the book and then helped steer it into a better book. Her coaching and guidance through the entire process have been invaluable.

Finally, I would like to thank Jennifer Ott, my wife and partner in crime for the past 25 years. In addition to tolerating (and tempering) my obsessive work habits, she also brought her eye as a designer and former editor to the book and greatly improved it, just as she has greatly improved me over the years.

THE NEW MIND READERS

CHAPTER 1

THINKING ON 20 WATTS
The Ultimate Scientific Challenge

Understanding how the brain works is almost certainly the most challenging scientific problem of our time. How can three pounds of tissue perform mental feats that outstrip the ability of the world's most powerful computers while consuming less energy than a dim lightbulb? Answering this question is the goal of neuroscientists, who study it at many different levels. Much of our current knowledge about how the brain works comes from studying other species, ranging from worms or fruit flies up to mammals like mice, rats, and monkeys. While this research has given us many important insights, most of us ultimately want to understand how the human brain works, and there are many aspects of the human mind that simply can't be studied in nonhuman animals: if we want to understand how humans think, we need to study humans.

This book will tell the story of how a set of new technologies has given us the ability to study how the human brain works in greater detail than ever before. These tools are known as *neuroimaging* methods, because they allow us to create images of the human brain that show us what it is made of (which we refer to as its *structure*) and what it is doing (which we refer to as its *function*). One tool in particular has revolutionized our ability to image the brain: magnetic resonance imaging (MRI). This incredibly versatile technique has provided neuroscientists with the ability to safely watch the human brain in action, which has allowed us to understand how the brain accomplishes many psychological functions. In some cases MRI can even allow us to decode what people are experiencing or thinking about by looking at their brain activity when they are performing a task or simply resting—what some audaciously call "mind reading" but what is more accurately known as decoding. And the power of MRI is not limited to studying the brain only during a fixed point in time. MRI has also shown us how experiences change the brain, and how individual human brains change over time from childhood to old age. It has shown us that all human brains follow the same general plan, but there are also many differences between people, and these studies have given insights into the brain dysfunctions that lead to mental illness. In doing so, MRI has raised many new questions that go beyond science, ultimately addressing some of the fundamental questions about how we view ourselves as humans. If thinking is just a biological function that we can visualize with MRI, then what becomes of the mystery of human consciousness? If decisions emerge from the computations of the brain, then in what sense are "we" responsible for our choices? Is addiction a "brain disease," a failure of self-control, or both? Should we worry about the ability of marketing researchers to use brain imaging to more effectively sell us their goods? It is these kinds of questions that we will grapple with after providing an overview of both the power and the limitations of neuroimaging.

What Is Neuroimaging?

When I use the term "neuroimaging" I am referring generally to a set of techniques that allow us to look at the human brain from the outside. There are a number of different ways to do this, but I will focus mostly on MRI because it has become the most widely used tool for brain imaging owing to its safety and its flexibility. Different kinds of MRI scans can be used to measure many different aspects of the brain, and we will roughly group them into what we call *structural* and *functional* MRI. Structural MRI measures different aspects of the makeup of brain tissue, such as how much water or fat is present in the tissue. Because different parts of the brain contain different amounts of these substances, they will show up on the MRI image as brighter or darker (see color plate 1). These aspects of the brain are very useful for detecting diseases of the brain, and for understanding differences in size and shape of different brain parts between people, but they don't tell us what the brain is *doing*—for that, we need to use functional MRI, or, as it is usually abbreviated, fMRI. fMRI came about when researchers discovered how to use MRI to detect the shadows of brain activity through its effects on the amount of oxygen in the blood. It is fMRI that provides the colorful images like the one shown on the right in color plate 1— in which parts of the brain seem to "light up." We will discuss the invention of fMRI and how it works in much more detail in chapter 2. First, we need to ask: What does "brain function" mean?

The Brain as a Computer

Each of the body's organs has evolved to serve a particular biological function: the heart pumps blood, the lungs oxygenate the blood, the digestive system extracts nutrients from food, and the kidneys filter waste products from the blood. What is the biological function of the brain? Whatever that function is, it's clearly very important—the brain accounts for only about 2% of the body's weight, but it uses about 20% of the energy consumed by the body.[1] If I had to come up with a simple label for what the function of the brain is, I would say: it processes information. Certainly not in the same way that your laptop or smartphone processes information, but nonetheless we can think of the brain as our body's central computing system, extracting information from the world and using it to choose how to act, with the goals of living long, prospering, and (most importantly for evolution) reproducing. The brain is not the body's only computer—for example, the gut has its own as well, known as the *enteric nervous system*, with about half a billion neurons—but it's certainly the most important when it comes to the things that make us uniquely human.

While it might make sense to call it a "computer," the brain is definitely not like most computers that we are familiar with in the world. Those silicon-based computers follow the recipe attributed to John von Neumann (one of the first true "computer scientists")—they are constructed from a large number of elements, which you can think of as microscopic switches, that behave in a highly reliable and consistent way. Anyone who ever experienced the "blue screen of death" on a personal computer knows what happens when one of these elements malfunctions; digital computers just aren't very resilient. These switches also operate very quickly. The computer that I am using to write this book has a clock speed of three gigahertz, which means that it can perform three billion operations every second, and thus those little switches have to be able to turn on and off very quickly. What's even more impressive is that those operations are happening largely "in serial," meaning that they are happening one at a time (or a few at a time in the case of most current computers).

How does the brain differ from a digital computer? To answer this, we need to understand how the cells in the brain process information, so it's time for a whirlwind tour of the physiology of the brain. The brain is made largely of two types of cells. Neurons are the cells that we have traditionally thought are essential to the brain's computing power. There is a second class of cells, known as glia, and these give the brain its structural scaffold and provide biological support to neurons. Until recently it was thought that glia were just supporting actors, but it's becoming more clear that they also play an important role in information processing.[2] However, throughout this book we will focus on the activity of neurons, since they are still the main type of cell studied by neuroscientists.

To understand how neurons work, let's trace the path of a signal from the world to the brain. When I brew my morning espresso and take a sniff, the smell that I experience starts with molecules from the coffee contacting my olfactory bulb, one of the only parts of the brain that is exposed directly to the outside world (right inside our noses). Those molecules hit a special type of neuron called an olfactory receptor, and cause changes in its cell membrane that increase the electrical charge within the cell. When this increase reaches a certain level, the cell suddenly changes its electrical properties, releasing what is called an *action potential*—a very sudden and large increase in its electrical charge. In common parlance, we say that the neuron "fired" or "spiked," because the change is so sudden. When this happens, the action potential travels down the length of the neuron and is ultimately transmitted to the other neurons that it connects to, and the cycle starts over—if the next neuron gets a strong enough input then it too will fire, and so on. If it's strong enough, the signal from the olfactory receptors will cause a cascade of activity between connected neurons that will ultimately reach my cerebral cortex, possibly triggering memories of the daily trips to the espresso bar on my first visit to Italy, or the desire for a piece of chocolate or pastry to go with it.

When we compare neurons in the brain to the central processing unit (CPU) in a digital computer, there is one important similarity. Most neurons behave in an "all-or-none" manner—just like a digital transistor, a neuron either fires or it doesn't, and all of its action potentials are basically the same in terms of their size and timing. This means that my olfactory neurons don't signal that the coffee smell is stronger by firing larger action potentials, but rather by firing more of them in succession or firing them more rapidly (figure 1.1). However, in almost every other way the brain computes very differently from a digital computer. First, brains are very slow compared to a digital computer—we measure the speed of computer operations in nanoseconds (billionths of a second), whereas the speed of firing of neurons is on the order of milliseconds (thousandths of a second). Second, individual neurons are noisy and unreliable. Of the many millions of neurons that are sensitive to the molecules in coffee, a different subset of them is going to fire each time I smell coffee. Third, brains process information in a highly parallel manner—rather than doing a few things at once very quickly like a CPU, the brain does lots of things at the same time, but does each one relatively slowly.

All of these features of the brain add up to a very different kind of computer, but that's a good thing. Most importantly, brains are resilient. If you drop your laptop and damage the motherboard, it's very unlikely that it will ever work again, and it certainly is not going to fix itself. The brain, on the other hand,

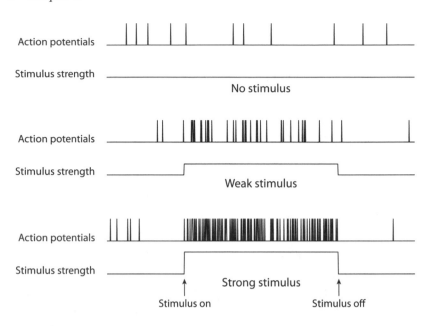

Figure 1.1. A simulated example of how an individual neuron responds to stimulation. The *top panel* shows action potentials as small spikes in the line, occurring even in the presence of no stimulation; most neurons fire randomly on occasion even when they are not stimulated. The *middle panel* shows the neuron's response to a weak stimulus, and the *bottom panel* shows the response to a strong stimulus. Note that the size of the action potential does not grow as the stimulus gets stronger; it simply fires more frequently.

is remarkably robust. Take the case of Lisa, whom I met when I was a postdoctoral fellow many years ago.[3] Lisa grew up as a relatively normal child, but around age 12 started suffering from severe epileptic seizures. Ultimately the seizures were so life-threatening and uncontrollable that at age 16 her doctors turned to a last-resort treatment known as hemispherectomy, which involves removing one entire hemisphere of the brain—fully half of her cerebral cortex (see figure 1.2). The seizures arose from her left hemisphere, which in most people is the side of the brain that is largely responsible for language function. Unfortunately this was the case for Lisa as well, and for the first year after her surgery she barely spoke at all. We studied her about six years after her surgery, at which point she had regained a remarkable amount of language function—far from normal for a 22-year-old, but nonetheless amazing given the fact

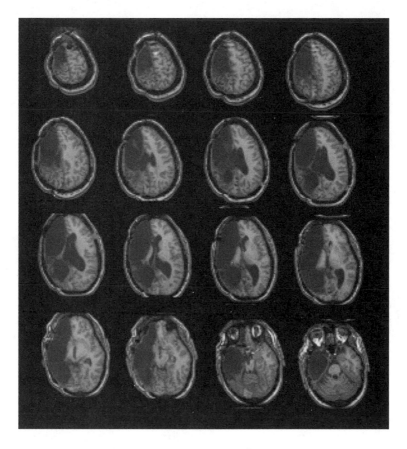

Figure 1.2. An MRI scan of Lisa's brain, showing that her left hemisphere is mostly missing, replaced by cerebrospinal fluid.

that her entire left hemisphere had been removed. This doesn't mean that half a brain is enough—after all, she was left with profound language problems—but it shows the amazing ability of the brain to recover from injury. We will return to the case of Lisa in chapter 5, where I discuss how neuroimaging allowed us to see how her brain had reorganized itself to allow her to speak and read.

There is another important way in which brains and digital computers differ. When you buy a personal computer, you have the choice of many different operating systems (such as Windows, Linux, or Mac OS) as well as a mind-numbing choice of software programs. This is because the software is fundamentally separate

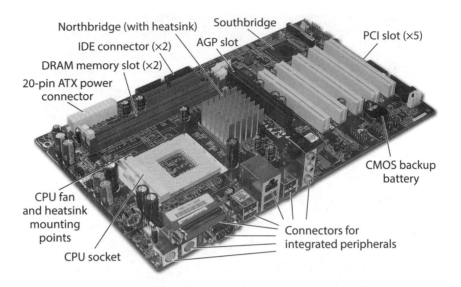

Figure 1.3. A computer motherboard with different parts labeled. Unlike the human brain, a computer is highly modular, with various parts that perform different specialized functions. By user: Moxfyre (Creative Commons license CC BY-SA 2.5 [https://creativecommons.org/licenses/by-sa/2.5]), via Wikimedia Commons.

from the hardware, for which we can also thank John von Neumann. In the brain, on the other hand, the hardware and software are inseparable; the "program" is stored in the connections between the neurons, and when we learn, it happens through changes in those connections. That is, the brain actually changes the configuration of its own hardware. We will discuss this further when we talk about brain plasticity, because it is these changes in connections between neurons that are crucial to the brain's ability to learn as well as its resilience to injury and disease.

Finally, it's important to understand how the architecture of the brain differs from the digital computers that we are familiar with, where by "architecture" I mean how the different parts function as part of the whole. Digital computers are built in a modular fashion, meaning that different parts have different specialized functions. Figure 1.3 shows the motherboard of a modern computer, with many of the different sections labeled. There are different parts of the motherboard dedicated to sound,

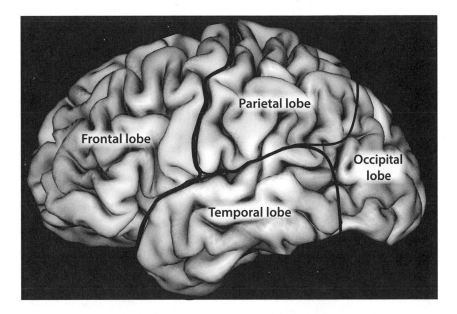

Figure 1.4. The lobes of the brain, outlined on an image of the surface of my brain that was generated from MRI data.

networking, memory, and the CPU, along with many other functions. We can tell that this system is modular in two ways. First, for many of the parts we can simply remove and replace them with a different version, as long as they are compatible. If a faster CPU or better video card comes along, I simply install it and start the computer up again, and with any luck it will just work. Second, damage to one of the parts will often have very specific effects; if I were to carefully damage the network chip (making sure not to damage any other parts), the sound functions of the computer should still work, and vice versa.

The non-neuroscientist could be forgiven for thinking that brains are also modular. After all, we regularly hear stories about neuroimaging studies that talk about "the reward center" or "the face area" in the brain. As I will discuss further below, there is a grain of truth in these stories, which is that functions are localized to some degree in the brain. People who have a stroke in their left prefrontal lobe are much more likely to have a language problem, whereas those who have a stroke that damages their right parietal lobe (see figure 1.4) are more likely to have difficulties with spatial awareness and attention. However, neuroscientists now realize that no single brain area does its work alone—there is no analog to the sound chip or video card in the brain. You should instead think of the brain more like a construction team. There are lots of specialist subcontractors (experts in perceiving speech, or finding locations in space, or predicting how another person will behave), along with a horde of general contractors who try to keep everything on track. However, no one of these individuals can build the building alone—it's the combination of all of them working together that creates the final product. As we will discuss in chapter 3, studies of how different regions of the brain communicate with one another have given us many new insights into how the brain functions as an integrated network rather than just a collection of specialists.

What Does the Brain Compute?

Now that we know a bit about how the brain computes, we are still left with a question: What return do we get on our outsized investment of energy in the brain? The short answer is that we get the ability to adapt. Just as humans can thrive on an amazing range of diets (from Inuits living on seal meat and whale blubber to Pacific islanders eating a diet of starchy vegetables like taro root), so too can the human brain adapt to a very wide range of cognitive environments and challenges. There are many organisms in the world that are adapted to function well within very specific niches, but few can function in as broad a range of environments as humans.

We think that one of the keys to the adaptive nature of the brain is its ability to build a predictive model of the world as it unfolds. It's easy to take for granted just how many predictions we are constantly making about the world around us. As I walk down the sidewalk, there is no guarantee that the concrete won't suddenly become liquid and swallow me, but I don't think twice as I take that next step; I assume that past experience is a good guide for the future. Similarly, when I ask one of my colleagues for advice on a problem, I assume that he or she will respond to me in English prose and not by singing operatically in Italian. Neither of these is guaranteed, but they are both pretty good bets. It's only when our predictions are violated that we realize just how beholden we are to them.

A large body of research in the past few decades has shown that the brain is constantly making predictions about the world, and updating those predictions when they are wrong. In fact, it's these violations that are at the center of learning; if we are behaving perfectly and the world abides by our expectations exactly, then why change anything? The neurotransmitter dopamine is one of the keys that relates learning to prediction errors. Dopamine is a different kind of neurotransmitter than the ones that send signals between specific neurons. Instead, we call it a "neuromodulator" because it changes the way that other neurons act, rather than causing them to fire directly. The dopamine neurons in the brain, which reside very deep in the middle of the brain, signal whenever something happens in the world that is unexpected; this could be something novel (imagine a loud noise in the library), or something that violates our expectations, either in a good way (such as finding a $100 bill on the sidewalk) or a bad way (such as finding out that your paycheck was smaller than expected). Dopamine neurons are constantly telling the rest of the brain how good the world is in comparison to our predictions, turning up their activity when the world exceeds our expectations and turning down when the world disappoints us. One of the major successes in neuroscience has been the development of a theory that links the role of dopamine in coding of "prediction errors" to our ability to adaptively improve our predictions, through a process called "reinforcement learning." We will discuss this in more detail in chapter 7 in the context of how the brain makes decisions.

From Brain to Mind

The brain is something that we can all get our heads around— it's a piece of tissue that we can see, measure, and if we are neurosurgeons, touch. But what is the mind? We all intuitively know what it feels like to have a mind, but what is it made of? The historical answer to this question was that the mind is on another plane. The philosopher Descartes famously argued that the mind (by which he really meant "soul") makes contact with the physical world in the pineal gland, a small structure in the middle of the brain. But how could something nonphysical interact with the physical world? That's the question that has perpetually dogged the concept of dualism—the idea that the mind is not of the physical world.

Dualism has not fared well in the age of neuroscience. The more we learn about the brain's workings, the clearer it is that our minds really are one and the same with our brains. An amazing example of this comes from research using electrical brain stimulation. It's common for researchers to stimulate and record from the brains of nonhuman animals, but those animals can't directly tell us about their experiences. In rare cases, however, researchers are able to examine the effects of direct electrical stimulation in the human brain. When someone suffers from severe epilepsy, surgeons will sometimes try to remove the part of the brain from which the seizures arise—in the case of Lisa this involved an entire hemisphere, but often a small part can be removed, which has a much less detrimental impact on the person's function. The goal is to remove only the part of the brain that causes the seizure, but it's usually not possible to tell which part that is from the outside, so surgeons will sometimes implant electrodes into the person's brain and then record from those electrodes for an extended period of a week or more. During this time, the person is sitting in the hospital with a bandaged head, waiting for the next seizure, and is often willing to participate in research to help alleviate the boredom.

One of the things that researchers can do is to stimulate the brain by running very small amounts of electrical current through the electrodes in the patient's brain. This is not enough to cause a seizure or damage the brain, but it is enough to change the patient's conscious experience in radical and sometimes bizarre ways. My colleagues Josef Parvizi and Kalanit Grill-Spector did an experiment in one such patient where they stimulated a part of the brain involved in face processing. The video of the patient's experience, published along with the paper, shows how a tiny bit of electricity can alter our experience:

> PARVIZI: Just look at my face and tell me what happens when I do this, alright? 1, 2, 3. [No stimulation is given.]
> PATIENT: Nothing.
> PARVIZI: I'm going to do it one more time, look at my face. 1, 2, 3. [Delivers 4 milliamp stimulation to face-sensitive area.]
> PATIENT: You just turned into somebody else. Your face metamorphosed. Your nose got saggy, went to the left. You almost looked like somebody I'd seen before, but somebody different. That was a trip.[4]

It is also possible to stimulate the brain in humans without surgery, though the stimulation is much less specific, using a technique called transcranial magnetic stimulation, or TMS for short. TMS involves putting a powerful electromagnetic coil up against the skull, and pulsing electricity through it for a very brief period (less than one thousandth of a second). When the electromagnet is pulsed, it causes a rapid change in the magnetic field underneath the coil, including in the brain tissue just below. Remember that neurons are conductors of electrical current, like little biological wires. We know from physics that a changing magnetic field will induce an electrical current in a conductor, and this is what happens underneath the coil: electrical activity is induced in the neurons below. If the pulses are strong enough they can actually cause a seizure, but researchers use much weaker pulses, which can alter brain activity without the risk of a dangerous seizure. Depending on how the pulses are applied, they can either stimulate or deactivate the neurons underneath the coil.

I experienced TMS firsthand in the 1990s when I volunteered for an experiment being run by my fellow postdoctoral researcher John Desmond. John wanted to use TMS to examine whether disruption of a specific brain area in the parietal lobe would affect the ability to hold information in mind, which we call working memory. In the experiment, I was shown a set of letters and had to hold them in memory, so that after a few seconds I could say which letters had been shown. On some of these trials he would zap my parietal lobe with TMS, hoping to scramble the activity there and disrupt my memory. I don't think the TMS had a very strong effect on my memory, but it did have another very striking effect: every time he zapped me, I experienced a strong metallic taste on half of my tongue. It turns out that the TMS pulse was probably also stimulating nerves in my face that are involved in the perception of taste, and that's why I had this experience. This highlights one of the challenges with using TMS to study the brain—its effects are fairly widespread, at best targeting an area roughly the volume of a golf ball, and often stimulating nerves or muscles outside the brain as well.

Studying the Mind

I have just made the argument that the brain and the mind are identical, which might lead you to think that studying the mind and studying the brain are the same thing, but that's not quite right. We call people who study the brain "neuroscientists" and those who study the mind "psychologists." In particular I am referring to experimental psychologists, who use experimental methods to try to understand how the mind works by testing hypotheses about how people will behave in certain situations. This is the field that I initially trained in; it was only after I came to Stanford in 1995 as a postdoctoral fellow that I began to use neuroimaging to study the brain.

A nice example of this kind of experimental psychology research comes from a study by the psychologists Henry "Roddy" Roediger and Jeff Karpicke of Washington University, who examined how we can best learn and remember new information.[5] In their studies, the experimental participants are presented with short paragraphs about various topics, such as sea otters or the sun. Participants were split into three groups: one group was told to read the paragraph four times (without getting tested at all), another was told to read it three times and was then tested once on the content, and the third was allowed to read it once, and was then tested on it several times. All of the groups were then asked how well they thought they had learned the materials, and the results were clear that the members of the first group (who had read the passage four times) were much more confident in their knowledge. The three groups were also tested on their knowledge of the material in the paragraph, either five minutes or one week later. The results of the immediate test were in line with the subjects' own predictions: memory was better for the people who had read the passage multiple times. However, a week later the results were strikingly opposite: The people who had been so confident immediately after learning had forgotten the most, and the people who had only read the paragraph once now had the best memories. Based on this research, the researchers proposed a theory that says that bringing information back from memory is actually one of the most powerful ways to cement the information into memory for the long term.

What's essential to note is that even though the research by Roediger and Karpicke is studying the workings of the brain, the study didn't measure the brain directly and the paper itself never mentions the brain. One can study the workings of the brain by measuring behavior, without actually looking at the brain itself. However, most psychology researchers now believe that the best way to understand the mind is to study both behavior and the brain simultaneously. This is the fundamental idea of the field known as *cognitive neuroscience*, of which I consider myself a member.

Cognitive Neuroscience before Neuroimaging

The focus of this book is on neuroimaging, which today is the most important tool of cognitive neuroscience. However, the field itself existed well before the advent of neuroimaging. Lore has it that the name "cognitive neuroscience" was hatched by Michael Gazzaniga and George Miller while they shared a taxi in the late 1970s. Miller was a famous experimental psychologist, perhaps best known for his 1956 paper titled "The Magical Number Seven, plus or minus Two," which pointed out that humans are limited to processing a small amount of information at once (usually about seven things) across many different domains. Gazzaniga, who most would consider to be the father of cognitive neuroscience, was famous for his studies of "split-brain" patients, which had shown how the two hemispheres of the brain can act independently. The field that these two researchers envisioned would combine psychology and neuroscience research approaches to provide a better understanding of how the brain gives rise to the mind.

Before neuroimaging, the only way to understand human brain function was to study people with brain damage and examine how specific damage leads to specific cognitive problems. This method first took hold in the nineteenth century, when European neurologists like Paul Broca and Carl Wernicke examined the postmortem brains of patients who had suffered from stroke, and noted that the location of the stroke corresponded with different types of language impairments. This is, in a sense, relying on Nature to do our experiments for us. However, Nature is an unreliable lab partner: Strokes can be large and messy, and results are often difficult to interpret for this reason. However, in rare cases the natural experiment can be much more specific. A fascinating example comes from a small set of individuals with a disease called Urbach–Wiethe syndrome, who have been studied by Ralph Adolphs from Caltech and his colleagues for a number of years. This is a disease that primarily affects the skin, but it also has a very specific effect on the brain, causing degeneration in a part of the brain called the amygdala, which has long been associated with emotion and fear. These patients have normal intelligence and mostly normal cognitive function, but they do show a very specific deficit: they largely do not experience fear. In one study, these researchers exposed an Urback–Wiethe patient (known by her initials, "S.M.") to stimuli that would make most of us shiver: live snakes and spiders, a haunted house, and clips from scary films like *The Blair Witch Project* and *The Shining*.[6] None of these things fazed her; in fact, the researchers reported that at the haunted house, "She reacted to the monsters by smiling, laughing, or trying to talk to them."[7] The only kind of fear that has been identified in these individuals is the fear of suffocation. This kind of research provides us with important clues about the brain systems that are involved in experiencing fear, and many other psychological functions have also been studied in this way, using patients with different kinds of lesions or brain disorders. Lesion studies still play a critical role in cognitive neuroscience because they allow us to ask a specific question: Is a particular brain region necessary for a particular cognitive function? Neuroimaging can't answer this question—sometimes brain areas can be active when a person does a task, but lesions to that area don't actually impair his or her ability to perform the task.

The Unlikely Success of fMRI

Since its invention in the early 1990s (which is discussed in much more detail in the next chapter), fMRI has overtaken all other methods in cognitive neuroscience, including lesion studies and other neuroimaging methods. However, in retrospect it's amazing that fMRI works at all. Its success relies upon a set of chemical and biological dominoes that all had to fall into place for it to have any chance of working, almost as if nature conspired to help make it just a bit easier for us to understand how the brain works (though just the tiniest little bit).

The first biological fact that makes fMRI possible is that the firing of neurons is relatively localized across the brain. Take the portion of the brain that processes visual information, which neuroscientists very creatively call the "visual cortex." Within this part of the brain, different sections respond to information coming from different parts of the visual world. Another part of the brain, in the temporal lobe (the auditory cortex), responds to sounds, and yet another (the motor cortex) makes my fingers move as I type words on the page. Different parts of the brain seem to do different things (that is, there is some degree of modularity, as we discussed above), and as we will see it is this localization of function that will ultimately allow us to decode what a person is doing or thinking of simply by looking at brain activity—the concept of decoding that I introduced earlier. It's possible to imagine that evolution could have constructed the brain very differently, with every function involving every part of the brain equally. In fact, until the middle of the twentieth century some very famous neuroscientists (such as Karl Lashley) believed that this was the case. However, the demonstration of clear effects of specific brain lesions on specific functions finally convinced the field that function is localized in the brain, at least to some degree.

Another aspect of the brain that makes fMRI possible is that brains are organized in a relatively similar way across individuals. Every human (and in fact nearly every mammal, except for monotremes such as the duck-billed platypus or spiny anteater) has a visual cortex that sits at the back of the brain, receives input from the eyes, and shows activity that is related to vision. Similarly, most mammals have a motor strip at the rear of the frontal lobe that controls the hands, whiskers, paws, or paddles. Again, we might imagine that evolution could have given us a random, haphazard organization of brain areas that varies from one individual to another, like spots on a calico cat. In this case, it would be very difficult to combine neuroimaging data across individuals, which we often need to do in order to gain statistical power through averaging. We also would not be able to compare the results with those from animal research, which can give us better insight into exactly what is happening in a particular area. Instead, research using nonhuman animals has provided important validation for results from fMRI research. The alignment across people is far from perfect, but it's good enough that we can warp together the brains of different people in order to analyze them as a group.

A third crucial biological fact is that the firing of neurons results in changes in blood flow that happen in a localized fashion as well. When neurons become active in a particular part of the brain, blood flow increases within the very close vicinity of those neurons (though we don't yet fully understand how this works). Without such tight localization, we would be able to see changes in blood flow but wouldn't be able to tie them closely to the neurons that caused them.

The final biological domino is the fact that this blood flow response to the area of active neurons is, in an important sense, an overreaction—at least with regard to oxygen. Blood brings with it a number of important things that neurons need, two of the most important being glucose and oxygen. What we know is that the brain seems to deliver about the right amount of glucose to make up for the energy used by the neurons when they fire, but it sends *too much* oxygen relative to the small amount that is used by neurons. The details of exactly how this works are still the fodder for spirited academic arguments, but what we know for sure is that it is this overflow of oxygenated blood that lets us detect the activity of neurons using fMRI.

The chemical fact that makes fMRI possible was discovered by the Nobel Prize-winning chemist Linus Pauling in the 1930s. He was studying the magnetic properties of the hemoglobin molecule, which is the molecule that carries oxygen in the blood. What he discovered was that oxygenated hemoglobin (which is what makes fresh blood red) was not magnetic, but deoxygenated hemoglobin was "paramagnetic." A paramagnetic substance is not a magnet itself, but it takes on magnetic properties in the presence of a magnetic field. Think of a paper clip, which is not magnetic on its own, but when put next to a bar magnet will become magnetic. The invention of fMRI took advantage of the relationship between oxygen level and the magnetic characteristics of blood, by developing particular ways to use the MRI scanner to detect these differences.

What Can't Neuroimaging Tell Us?

While fMRI has shown itself to be incredibly powerful, it has also been used in ways that go beyond what it can actually tell us, which was illustrated well in an event from 2007. On November 11 of that year, an op-ed piece titled "This Is Your Brain on Politics" was published in the *New York Times*.[8] The authors, well-known neuroscientists and political scientists, reported results from a study in which they used fMRI to measure brain activity while so-called "swing voters" viewed video clips of candidates in the then-ongoing US presidential primaries. Based on these data, they drew a number of broad conclusions about the state of the electorate, which were based on the brain areas that were active while viewing the videos. One of the claims in the op-ed was that:

> Emotions about Hillary Clinton are mixed. Voters who rated Mrs. Clinton unfavorably on their questionnaire appeared not entirely comfortable with their assessment. When viewing

images of her, these voters exhibited significant activity in the anterior cingulate cortex, an emotional center of the brain that is aroused when a person feels compelled to act in two different ways but must choose one. It looked as if they were battling unacknowledged impulses to like Mrs. Clinton. Subjects who rated her more favorably, in contrast, showed very little activity in this brain area when they viewed pictures of her.

Here was the verdict on Barack Obama:

Mr. Obama was rated relatively high on the pre-scan question-naire, yet both men and women exhibited less brain activity while viewing the pre-video set of still pictures of Mr. Obama than they did while looking at any of the other candidates. Among the male subjects, the video of Mr. Obama provoked increased activity in some regions of the brain associated with positive feeling, but in women it elicited little change.

As I read this piece, my blood began to boil. My research has focused on what kinds of things we can and cannot learn from neuroimaging data, and one of the clearest conclusions to come from this work is that activity in a particular region in the brain *cannot* tell us on its own whether a person is experiencing fear, reward, or any other psychological state. In fact, when people claim that activation in a particular brain area signals something like fear or reward, they are committing a basic logical fallacy, which is now referred to commonly as *reverse inference*. My ultimate fear was that the kind of fast-and-loose interpretation of fMRI data seen in the *New York Times* op-ed would lead readers to think erroneously that this kind of reasoning was acceptable, and would also lead other scientists to ridicule our field.

What's the problem with reverse inference? Take the example of a fever. If we see that our child has a fever, we can't really tell what particular disease he or she has, because there are so many different diseases that cause a fever (flu, pneumonia, and bacterial infections, just to name a few). On the other hand, if we see a round red rash with raised bumps, we can be fairly sure that it is caused by ringworm, because there are few other diseases that cause such a specific symptom. When we are interpreting

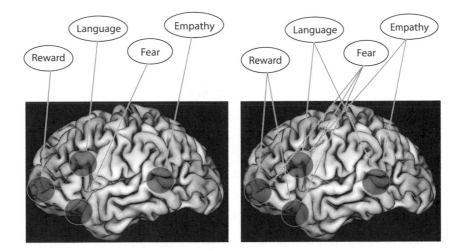

Figure 1.5. Can you infer cognitive function from areas of brain activation? If there was a one-to-one mapping between brain areas and cognitive functions, as shown in the *left panel*, then reverse inference based on activation in those areas would be possible—activation in the amygdala would imply fear, and activation in the ventromedial prefrontal cortex would imply reward. However, the brain is actually organized more like the *right panel*—any mental function involves a combination of many different brain regions, that are combined in different ways to support different mental functions.

brain activation, we need to ask the analogous question: How many different psychological processes could have caused the activation? If we knew, for example, that mental conflict was the only thing that causes the anterior cingulate cortex to be active, then we would be fairly safe in concluding from anterior cingulate activity that the person is experiencing conflict when viewing images of Hillary Clinton. On the other hand, if many different things can cause the region to be active, then we can't safely draw that conclusion. Figure 1.5 shows an example of each of these two different cases. Work that I published in 2006 showed that activity of individual brain regions was not very specific for different psychological functions (that is, it's more like a fever than a round rash), and thus that this kind of simple reverse inference is problematic.[9] The anterior cingulate cortex is a prime example of this. When we looked across many thousands of published neuroimaging studies in a later study, we found that this area was active in about one-quarter of all those studies, which involved many different types of cognitive tasks.[10] This means that we cannot tell very much at all about what a person is doing from the fact that the anterior cingulate cortex was active.

Throughout this book, I will return to the fact that there is no simple one-to-one mapping between psychological states and activity in specific brain areas. As we will see, it is possible sometimes to decode the contents of a person's mind using fMRI, but it requires sophisticated statistical analyses along with careful interpretation.

A Road Map for the Book

The goal of this chapter was to give some background in the kinds of questions that cognitive neuroscientists ask, as a prelude to discussing how we use neuroimaging to ask them. In the rest of the book I will lay out the story of how neuroimaging came to be, what it can and can't tell us, and where it is going.

The first section of the book focuses on the development of neuroimaging as a tool for studying the mind and brain. In chapter 2, I will describe how researchers in the 1980s began to use a method called positron emission tomography (PET) to study how brain activity related to mental function, building on a century of prior ideas about the relation between brain activity and blood flow. Their discoveries led to the development of functional magnetic resonance imaging (fMRI), which is now the dominant technique for measuring brain activity in humans, relying on an amazing confluence of biology, chemistry, and physics. In chapter 3, I discuss how fMRI grew from being a new technique to the most powerful tool in human neuroscience. We will see how it was validated as a measure of brain activity, and how it was used to ask specific questions about how the brain is organized. In chapter 4 we will dig more deeply into how fMRI has been used to decode the contents of the mind and to attempt to achieve "mind reading." In chapter 5, I will discuss how fMRI has shown us how experiences change the brain, and how individual human brains change over time.

The second section of the book will focus on the ways in which neuroimaging has begun to influence the world outside of the laboratory. In chapter 6, I discuss the ongoing attempts to use neuroimaging evidence in the courts, including the attempt to use fMRI for lie detection and why this is currently problematic. In chapter 7, I discuss the use of neuroscience tools to better understand how humans make choices and, ultimately, use them to find ways to sell us things more effectively, via the new field of consumer neuroscience. In chapter 8, I discuss how cognitive neuroscience has improved our understanding of mental illness, and discuss the ethical and social challenges of thinking of mental illnesses and addictions as "brain diseases."

Finally, in chapter 9 I discuss the future of fMRI and how its limitations may be addressed using other new methods.

CHAPTER 2

THE VISIBLE MIND

Having a brick dropped on one's head is not a usual path to a place in scientific history, but that is how it happened for Michele Bertino. In 1877, he was standing next to the bell tower in his small Italian village when a worker on the top of the tower dropped a brick about 40 feet, hitting him on the head. One might imagine that this would lead to certain death in the days before antibiotics and clean surgery, but Michele survived, albeit with one small catch: he was left with a hole in his skull, as this was before the days of reconstructive surgery. A couple of months after the accident, now back on his feet, Bertino went to visit a physician and professor at the University of Turin named Angelo Mosso. Mosso had developed a method to measure blood pressure, and he used it to measure the pulsations in Bertino's brain through the hole in his skull. He did many different measurements, but the ones that were really momentous for neuroscience explored what happened when Bertino engaged in thought. What Mosso found was that thinking caused the pulsations to increase specifically in Bertino's brain—the pulse at his wrist didn't change (figure 2.1). This led Mosso to the idea that the circulation of the brain responds to mental activity—an idea that was revolutionary in the 1880s, when he published his book *On the Circulation of the Blood in the Human Brain*.

Mosso's next experimental move was also a harbinger of things to come. He reasoned that if mental activity caused an increase in the amount of blood in the brain, then this should change the relative weight of the head compared with the rest of the body. To test this, he built a device he called the "human circulation

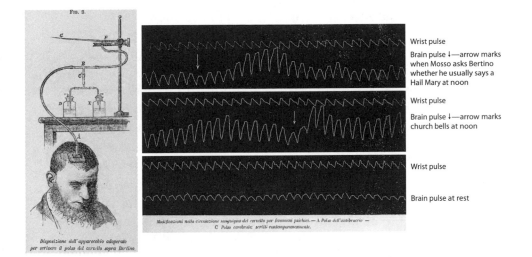

Figure 2.1. Angelo Mosso's measurement of the pulsations in Bertino's brain. Figure from Mosso's book (*left*) showing a schematic of the device that he used to measure the pulsations in Bertino's brain. Another figure from the book (*right*) showing examples of how the pulsations of Bertino's brain (the *bottom trace* in each graph) changed in response to different events, while the blood pressure in his arm (*top trace*) remained constant. Thanks to Pietro Pietrini for assistance with translation of the original Italian text.

balance," which was basically a bed that balanced around a point in the middle—though it was actually much more complicated than that, because he had to compensate for the movements in the bed caused by breathing. Once the experimental subjects had lain long enough for their blood to distribute throughout the body, he had them read materials, from the simple (a newspaper article) to the complex (philosophy or mathematics). What he saw was that as the subject engaged in more difficult mental activity, the balance tipped toward the head. These informal experiments provided some evidence but left many questions open. However, a modern-day re-creation of the circulation balance by David Field and Laura Inman of the University of Reading has provided evidence that confirms Mosso's claims.[1] These authors built a similar device, using a sensitive scale under the head to measure the movements of the table. In one test, they had subjects hold their breath; this results in an increase of carbon dioxide in the blood and a subsequent increase in blood flow to the brain. Looking at the traces of the scale they could see the weight at the head of the table slowly rise during the time that the breath was held (causing blood flow to increase), demonstrating that their balance could successfully measure blood flow to the brain. Then they tested whether mental stimulation could result in similar changes in blood flow. In one experimental condition subjects were presented with an auditory stimulus (music) for two seconds, while in another they both heard the auditory stimulus and saw a visual pattern generated from the music by the Windows Media Player. What the researchers found was that while the blood flow to the head did not increase appreciably for the auditory condition compared to rest (likely owing to a problem with their experimental design), blood flow did increase for the audio/visual condition. The results from Field and Inman confirmed that Mosso had indeed built a nineteenth-century version of a brain imaging system, though it was clearly quite crude because it measured the entire brain at once.

Mosso's findings had also inspired other researchers to study blood flow in the brain using animal models, including Charles Sherrington (who would win the Nobel Prize in 1932 for his discoveries about how neurons function) and C. S. Roy, both of whom were working at Cambridge University at the end of the nineteenth century. They set out to understand the factors that affected blood flow to the brain by studying the brains of dogs, cats, and rabbits. In many of their experiments they found that the pulsations in the brain were simply a reflection of changes in the blood pressure outside of the brain, but one experiment provided critical evidence about the control of blood flow in the brain. (Warning: the following is not so pleasant for animal lovers.) One dog was bled to death, and its brain was ground up and mixed in a saltwater solution. The idea was that this brain tissue would have in it the chemicals that are released when blood flow is low, which should cause an increase in blood flow in the brain of a healthy organism. This tissue was then injected into the bloodstream of another dog, while the blood flow in both its brain and its general circulation was measured. What they found was that the dog's brain expanded, indicative of increased blood flow, whereas the blood pressure outside the brain did not change. They concluded:

> These facts seem to us to indicate the existence of an automatic mechanism by which the blood-supply of any part of the cerebral tissue is varied in accordance with the activity of the chemical changes which underlie the functional action of that part. Bearing in mind that strong evidence exists of localisation of function in the brain, we are of opinion that an automatic mechanism, of the kind just referred to, is well fitted to provide for a local variation of the blood-supply in accordance with local variations of the functional activity.[2]

Little did Roy and Sherrington know that they had uncovered the mechanism that 100 years later would allow neuroscientists to see the mind in action at a level of detail that they could have scarcely imagined.

The Birth of Cognitive Neuroimaging

Despite the enthusiasm about the work of people like Mosso and Roy and Sherrington, the idea of measuring blood flow to understand the mind would not gain real steam until the 1980s, with the advent of a new technique called positron emission tomography (or PET for short).[3] PET takes advantage of the fact that when particular radioactive isotopes break down, they give off a particle known as a positron (also known as an antielectron). After it is emitted, the positron travels a short distance until it runs into an electron, which results in the emission of two high-energy photons traveling in exactly opposite directions through a process known as "annihilation." To create a PET image, the individual is injected with a radioactive tracer, and a ring of detectors arrayed around the head detects the emitted photons and keeps track of which locations in the ring were hit at the same time. The "tomography" part of PET refers to the method by which one can use the information about the paths taken through the brain by these photons in order to reconstruct an image of where the radioactive decay is happening most strongly.

PET scanners are now commonly used in hospitals, particularly for the detection of cancer. The reason that PET is useful for detecting cancer is that tumors tend to have higher blood flow and use more glucose compared with healthy tissue. If we inject a radioactive tracer that will be carried in the blood (such as radioactive water), then areas of increased blood flow should end up receiving more of the tracer, and thus should have more radioactive decay, which we will see on the PET scan. Similarly, if we attach a radioactive isotope to the glucose molecule, we can then image areas that have increased glucose usage. It turns out that both of these also occur when neurons are active, and PET can be used to image them, though most research has used radioactive oxygen because it decays much faster (about two minutes versus almost two hours for glucose labeled with radioactive fluorine).

The first concerted use of neuroimaging to study mental functions can be traced to a collaboration between two pioneering researchers from different fields, spurred by an unlikely matchmaker. One of them, Marcus Raichle, is a remarkable neuroscientist whose name will appear at multiple points throughout this book (see figure 2.2). Raichle was trained as a neurologist, served in the Air Force School of Aerospace Medicine during the Vietnam War, and then moved to Washington University in St. Louis in 1971, where he has been a member of the faculty ever since.[4] In the early 1980s he was witness to the invention of the PET scanner, which was largely driven by another young researcher at Washington University named Michael Phelps, who had been working alongside Raichle in the laboratory of Michel Ter-Pogossian. While Phelps and Ter-Pogossian developed the physical machinery for PET scanning (often referred to as a "camera"), Raichle and his group were responsible for developing some of the first techniques for analyzing PET imaging data.

The other half of the duo was a psychologist named Michael Posner, a pioneering cognitive psychologist from the University of Oregon. Posner's early work had used studies of reaction times to investigate many different aspects of cognition, such as attention, memory, and categorization. He later became

Figure 2.2. Dr. Marcus Raichle alongside one of the early positron emission tomography (PET) scanners at Washington University in St. Louis. Image courtesy of Marcus Raichle.

interested in learning about the brain systems that underlie these functions, and ended up joining the Washington University team as a collaborator. Posner also had a special connection to Raichle: his brother Jerry, a neurologist in New York, had been one of Raichle's supervisors in his early years. As Posner says in his autobiography: "In my 3 years in the Neurology and Neurosurgery Departments in St. Louis at the start of cognitive imaging, being Jerry's brother helped to overcome my connection to psychology and cognition, both regarded with suspicion in that department."[5] This quote highlights the sometimes uneasy nature of the merger between psychology, neuroscience, and imaging; many neuroscientsts and imaging scientists still harbor feelings that the mind is just too ephemeral to study scientifically.

Another person who gets no scientific credit, but nonetheless played a crucial role in enabling the birth of cognitive neuro-imaging, was the St. Louis aviation pioneer James S. McDonnell (known to his family and friends as "Mr. Mac"). McDonnell was a corporate titan in the aviation industry, having founded the McDonnell Aircraft Corporation and later McDonnell Douglas, but he had always had an interest in the mind as well as in the paranormal. According to McDonnell's son:

> In the area of human cognition he had a lifelong fascination with the workings of the brain and the mind. When he learned of the new positron emission tomography, developed by a team at Washington University, Mr. Mac immediately con-vened a meeting of Washington U's top researchers to discuss how neuronal activity underlies mental activity and behavior. He challenged the group to prepare a truly innovative research proposal for consideration by the McDonnell Foundation. By the time the proposal was presented in May 1980, Mr. Mac had already suffered the first of a series of strokes from which he died in August of that year. However, he rallied his faculties to further improve the proposal and made a $5 million grant to endow the Center for the Study of Higher Brain Function, which was one of his last acts before passing on to the next phase of existence (as he characteristically referred to death).[6]

Michael Posner describes a slightly different take on the situation:

> James S. McDonnell had wanted to develop an Institute that would study extrasensory perception, but the powers that be at Washington University were not going to do that. Instead, they agreed to a Center for Higher Brain Function. A psychologist who studied brain function was about as mystical as they want-ed to go, and Marc Raichle and his colleagues at Washington University recognized the importance of being able to use PET to illuminate questions of higher mental function.[7]

It was the funding from the McDonnell Foundation that allowed the collaboration between Raichle and Posner, along with their trainees Steven Petersen and Peter Fox, to flourish. In future years, the foundation would provide training support and seed funding to many individual researchers in the field of cognitive neuroscience, including myself.

The work by the Washington University group revolutionized how we study the relation between the mind and brain. One of the most important ideas that they utilized was the concept of the *subtraction method*, an idea first developed by the Dutch psychologist F. C. Donders in the nineteenth century, who had set out to understand how long different psychological processes take. To understand the subtraction method, imagine that we want to know how long it takes to process the meaning of a word. In order to estimate this, the research subject needs to engage in some kind of cognitive task that requires processing the meaning of words so that we can measure the time that it takes. Let's say that I ask the person to tell me as quickly as possible whether the word is the name of an animal or a vegetable, which clearly requires an understanding of the meaning of the word. I ask the subject to press one of two keys on a keyboard, depending on the answer. Let's say that I show the word "carrot" and it takes the person 800 milliseconds (which is 800 thousandths of a second) to press the correct key. We would not want to infer from this that processing the meaning of the word takes 800 milliseconds, because there are other aspects of doing the task that also take time, such as seeing the visual features of the word, recognizing the letters, and pressing the button. What Donders realized is that by comparing different psychological tasks, he could use differences in the reaction times between them to estimate how long different "stages" of mental processing require. For example, in order to estimate the amount of time needed to process the word's meaning, we might compare the reaction time on the animal/vegetable task to the time on another task where the person simply has to decide whether a string of letters is a word (such as "piglet") or not (such as "flxvbd"). The difference in reaction times between these different tasks would give us the estimate of how long it takes specifically to access the word's meaning, while "subtracting out" many of the other aspects of doing the task, such as the visual recognition of the letters and the pressing of the button to signal the decision.

The Washington University group took Donders's subtraction method and applied it to brain activity instead of reaction times. In 1988 they outlined their results in two papers published in the prestigious journals *Science* and *Nature*.[8] In one set of experiments they presented subjects with a set of tasks that were meant to isolate different aspects of language processing. The simplest one compared passively looking at words on a screen to simply staring at a "fixation" point (a plus sign) on the screen; the more complicated tasks involved reading the word aloud, or generating a verb that was meaningfully related to the noun on the screen. Subjects each participated in several scans, in which they were first injected with water labeled with radioactive oxygen, and then were told to perform one of the different tasks for 40 seconds, during which time radioactive emissions from their brains were measured using the PET scanner. These images show the overall blood flow in the brain during the task—while some of this blood flow is related to doing the task, most of it is not, and simply reflects the high metabolic needs of brain tissue. In order to determine which brain areas were specifically active in relation to different mental processes, they subtracted the images for the different tasks—the difference between the image collected during the verb generation task and the image collected during the reading task was thought to index the extra brain activity needed to access the meaning of the word in order to generate a related verb.

After these data were collected, there was another problem that had to be solved; fortunately, Raichle's team was already working on it. The challenge was to find a way to average together the data from multiple individuals, in order to overcome the noisiness of the data from each individual. Each person's brain is different in size and shape, and, in addition, each person's head will be in a slightly different location inside the PET scanner. Because of this, the data from all individuals have to be somehow aligned in a common reference frame so that they can be averaged together. To do this, Raichle and his team took advantage of a technique that had been developed by the French neurosurgeon Jean Talairach to align different brains together for the purposes of neurosurgery. Using an X-ray image that was collected at the same time as the PET scan, they moved and stretched the PET images (which are like a set of flat slices through the brain) until they matched the atlas that Talairach had developed, which resulted in all of the brains being roughly lined up with one another. Having done this, they were then able to average the data across the different individuals who had participated, and then compute statistics in order to find areas where they had good confidence that the activity was signal rather than noise.

Using these methods, the group identified a set of brain regions whose activity differed between specific pairs of cognitive tasks. Comparing the simplest word tasks (seeing or hearing words) to staring at a point on the screen identified areas of activity in the brain's visual cortex for seeing and the brain's auditory cortex for hearing; this was in no way a new discovery, but it gave everyone confidence that the method was working properly. The real surprise came when they examined the brain areas that were more active for the generation of verbs than for the reading aloud of words. The studies of brain-damaged patients that started with Carl Wernicke had suggested that the processing of meaning happened primarily in the temporal lobes of the brain (figure 2.3). But that's not what the PET data showed—instead, the areas that were more active for the processing of meaning were in two different parts of the frontal lobes. The same areas showed up regardless of whether the subjects read or heard the words that they processed. This finding started a decade-long search to understand the role of the prefrontal cortex in the processing of meaning, a search that I played a role in a decade later in one of my first neuroimaging studies. We now think that different parts of the prefrontal cortex play different roles in language processing, with a more forward area being crucial for the retrieval of the meaning of words; the meanings themselves are thought to be stored primarily in the temporal lobes. The results from Petersen and colleagues made a big splash, in part because they provided the first glimpse into cognitive processes in the brain, but also because they

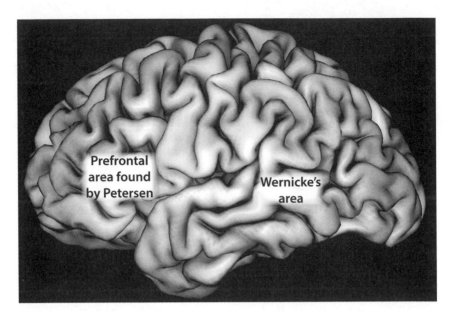

Figure 2.3. The location of Wernicke's area (where researchers expected to find activity related to the processing of meaning) and the left prefrontal areas that Petersen and his colleagues found to be active during processing of meaning.

showed how neuroimaging could actually tell us something that we hadn't already learned from human lesion studies or from research on animals.

The excitement about the new PET results was tempered by some of the inherent limitations of PET imaging. First, it involved exposing subjects to radioactivity, which limited the amount of data that could be collected from any individual, and also meant that one couldn't scan children. Second, PET requires very expensive equipment and is not widely available. Third, PET has a resolution problem. It is now common to have digital cameras with more than 10 megapixels (that's 10 million pixels). The reason that more pixels are generally better is that they provide higher "spatial resolution"—that is, they let us distinguish smaller objects in the image. Whereas the letters in a far-off sign might be too blurry to distinguish with an older camera, the additional spatial resolution of a more modern camera can help better identify these smaller

Temporal resolution

Figure 2.4. Examples of the effects of spatial and temporal resolution on a photo. Spatial resolution reflects the sharpness of the image in space; a photo with low spatial resolution appears blurry all over. Temporal resolution reflects the sharpness of the image in time; things that are changing quickly in time will look blurry in a low temporal resolution image.

features (figure 2.4). The spatial resolution of PET is about one centimeter, which is much larger than most features of interest in the brain. For example, the cerebral cortex is made up of small units known as "cortical columns," which are sets of neurons that are tightly interconnected with one another and process similar kinds of information. In the part of the brain that is sensitive to visual motion, each column is sensitive to motion in a specific direction. These cortical columns are less than one millimeter in width, which means that we would need spatial resolution much higher than PET's in order to disentangle them.

In addition to taking images of brain activity that are sharp, we also want to take images that are fast; the speed with which we can take an image is known as "temporal resolution." Imagine taking a picture of two runners crossing the finish line in the 100-meter event in the Olympics. If the camera needs one second to take the image, then the runners will be a blur and it will be impossible to tell who crossed first. In order to capture a sharp picture of the runners crossing the finish line, we need to take the image very quickly: the camera used at the 2016 Olympics can take 10,000 images per second. PET fails badly on this account; it takes at least tens of seconds to collect enough data to create an image using PET, which means that at best PET can only give us a very blurry snapshot of brain activity.

A Magnetic Camera

In the 1980s, the up-and-coming technology in medical imaging was magnetic resonance imaging. The concepts behind MRI had been developed in the previous decade, but it wasn't until the 1980s that MRI scanners started showing up in hospitals around the world in significant numbers. MRI is quite an amazing tool, mostly because of the flexibility with which it can be used, as well as its safety. With PET, the only way to image a new type of biological process is for a chemist to attach a radioactive tracer to a molecule that is involved in that process. Not only is this technically challenging work, but the resulting chemicals also have to be tested for safety before they can ever be injected into a human. MRI, on the other hand, doesn't involve ionizing radiation and doesn't require anything to be injected into the human (though sometimes so-called "contrast agents" are injected to make certain types of tissue show up more clearly)—in general, examining a different aspect of the tissue with MRI simply requires writing a different computer program (called a "pulse sequence") for the MRI scanner. The lack of radiation also means that an individual can be scanned many times safely—a fact that will become critical for a study that I will discuss in chapter 5 in which I was scanned more than 100 times over the course of two years.

MRI is based on the concept of "nuclear magnetic resonance," which was discovered in the middle of the twentieth century; the word "nuclear" was dropped from "magnetic resonance imaging," in part over concerns that the term would frighten patients because of its association with ionizing radiation and atomic weapons. The "magnetic" part of MRI comes from the fact that it measures a particular kind of resonance that happens when material is placed in a magnetic field (which is nuclear magnetic resonance). When a material is not exposed to a magnetic field, the nuclei of the atoms will be spinning in random directions—imagine spinning tops in zero gravity. When we put material in a magnetic field, some of the nuclei will become aligned with the magnetic field. They are now pointing in the same direction, but their spinning is not synchronized— imagine a set of identical tops spinning on a table, each started at a different time. The final trick that we can play is to apply a pulse of energy—in this case radio frequency (RF) energy—into the material, which causes the spinning nuclei to line up with one another. This is what we mean by "resonance." Just as we can hear the sound waves coming from a piano as it resonates, with the right kind of antenna we can measure the electromagnetic signals coming from the material that has been excited by an RF pulse.

The remaining aspect of MRI that we need to explain is the "imaging" part. The process I describe above gives us a signal coming back through the antenna, but it doesn't tell us where in the tissue it's coming from; we just get one signal for the entire tissue. In order to make an image that shows us what is happening at different locations within the tissue, we need to do a couple more things. First, we can excite the nuclei in just part of the tissue. To do this, we can take advantage of the fact that the speed at which a nucleus spins is determined by the strength of the magnetic field, and that our excitation only affects nuclei that are spinning at a particular speed. An MRI scanner is built in such a way that the magnetic field can be varied gradually along the length of the tube (which we call the "bore"), being higher at one end and lower at the other. This means that the speed at which the nuclei spin differs slightly in different locations along the bore, and thus we can excite a specific slice of tissue using a pulse of RF energy that matches the speed at that location. That gives us a slice, but it doesn't tell us where the signals are coming from within that slice. To do that, we need to add some information to the signal so that we can decode where in space the different signals are coming from, which we call "encoding." There are many different ways to do this, but all of them involve tagging the signals in particular locations with particular phases and frequencies. Then we can use a mathematical technique known as the Fourier transform to unmix the signals coming from the different locations and "reconstruct" those into an image.

The magnetic fields used in MRI scanners are very strong— strong enough to rip a pair of pliers from one's hand from several feet away. Magnetic fields are measured in units called teslas; a small bar magnet has about 0.01 tesla, and the Earth's magnetic field is roughly 0.00005 tesla. Most standard clinical MRI scanners have a magnetic field strength of 1.5 tesla, while scanners used for research can be anywhere from 3 to 11 tesla. Despite these very strong magnetic fields, it is thought that MRI is generally safe for humans, unless they have certain kinds of medical implants or devices that might be affected by the magnetic field. However, the magnetic fields can have visible effects, especially at very high field strengths. One interesting side effect of high magnetic fields (7 tesla and higher) is that moving one's head too quickly in the magnetic field can cause an experience something like motion sickness. This happens because microscopic crystals in the inner ears called *otoconia*, which are involved in sensing movement of the head, experience a slight force from the magnetic field. For this reason, the bed must move very slowly when putting a person into a high field MRI scanner, so that he or she doesn't experience these effects.

The images that were available from the early MRI scanners were focused on aspects of the brain related to the structure of the tissue, such as how much water or fat is present in it (see color plate 1). These images were very useful for identifying brain diseases (such as multiple sclerosis or brain tumors), but they didn't tell us anything about what the brain was doing at any point in time (that is, its function). The development of *functional* magnetic resonance imaging (fMRI) was the technological breakthrough that allowed researchers to start to look at what the brain is doing, not just what it is made of.

"I Will Make the Visual Cortex Light Up"

Scientific discovery relies upon a lot of factors, including hard work, persistence, luck, and, perhaps above all, timing. Jack Belliveau had all these, but also much more. Nearly everyone who knew Jack describes him using terms like "larger than life" and "boundless enthusiasm."[9] Jack arrived at Massachusetts General Hospital (MGH) as a graduate student in biophysics at Harvard Medical School in the early 1980s, intent on studying the brain using MRI. He landed in an environment that was buzzing with excitement about the application of MRI to many different body parts and diseases, but he seems to have been the one who came up with the idea of studying the mind using MRI, apparently inspired by his dream of recording and storing a copy of the conscious mind so that it could be replayed later. However, his path to discovery was a convoluted one, to say the least. He started out working in the laboratory of a biophysicist named Eric Fossel, but was apparently kicked out of the lab for breaking a piece of equipment. This was almost certainly a lucky event for two reasons: first because he landed in Tom Brady's lab where he was able to pursue his dream of imaging the brain, and second because Fossel would later be found guilty of falsifying scientific data and be fired by Harvard Medical School. In Brady's lab Jack met another young researcher named Bruce Rosen, who would be his mentor and collaborator for many years and would go on to lead the MGH Nuclear Magnetic Resonance (MGH-NMR) Center, now known as the Martinos Center for Biomedical Imaging. I worked at the MGH-NMR Center from 1999 to 2002; I didn't see Jack very much during my time there, but his legend filled the halls, as it continues to even after his untimely death in 2014.[10]

The research that led to the development of fMRI started with a set of researchers studying the liver (who were known to their colleagues as "liver breaths"). They had developed a way to image the liver fast enough to see changes in the MRI signal that occurred when an animal was injected with a substance known as a "contrast agent" (because it causes differences in the brightness of the image, which we generally refer to as "contrast"). However, while their knowledge of physics had led them to expect that the images should become brighter when the contrast was injected, they saw that it actually got darker. A young German neurologist at MGH named Arno Villringer set his sights on understanding this conundrum. He and his colleagues discovered that the decrease in brightness of the liver after the injection was related to an effect of the contrast agent on something called "magnetic susceptibility"—that is, how much the tissue becomes magnetized when placed in a magnetic field. This was a key discovery in figuring out how to use MRI to study brain function. Jack Belliveau set out to examine this using a new experimental 1.5-tesla MRI scanner that had just been delivered to MGH, which was able to collect images much more quickly than the standard MRI scanners in the hospital, using a technique called "echo-planar imaging" (EPI). This scanner was big enough to fit a human, but Jack started by testing out his ideas in dogs because he could control their physiology more closely. He first duplicated Villringer's earlier experiments (done in rats) and showed that, as expected, the brightness of the image went down when he injected the contrast agent. He then tried to relate this to blood flow, by looking at what happened when he caused the level of carbon dioxide (CO_2) in the blood to increase. He saw that the levels of signal in the brain caused by the contrast agent went up in lockstep with the amount of CO_2 in the blood, which told him that he was measuring blood flow.

So far they had imaged blood flow, but had not related it directly to brain activity. In 1990 Jack Belliveau and his colleague David Kennedy traveled to the Dartmouth Summer Institute for Cognitive Neuroscience, and happened to have a discussion with Peter Fox from Washington University, who was part of Raichle's team and one of the coauthors of the landmark 1988 paper on cognitive imaging using PET. The story goes that Jack borrowed a set of flickering-light goggles that had been used by Fox and Raichle in their PET studies to stimulate the visual system, and then proceeded to use them to try to replicate those studies using MRI. Using the experimental MRI scanner sitting in Bay 3 at the MGH-NMR Center in Charlestown, they measured MRI signals from the visual cortex by injecting the contrast agent while the goggles were either on or off. What they saw was that the signal went down whenever the contrast agent was injected, but it went down more when the goggles were flickering, in the same part of the brain where the earlier PET work had seen a response. This was the first time that brain function had been directly measured using MRI, and the resulting paper was published in *Science* in November 1991, with an image of the results as its cover image (see color plate 2).

The BOLD Road to fMRI

The landmark work by Belliveau and his colleagues had shown that MRI could be used to measure brain activity, but his approach had a major drawback: it required injection of a contrast agent. In addition to being inconvenient, these contrast agents can cause side effects, including very rare but potentially deadly side effects in people with existing kidney or liver problems. In the wake of Belliveau's publication, researchers scrambled to develop a method to image brain activity that did not require injection of a contrast agent. By late 1991, three different groups were working to develop fMRI without injected contrast.

The first group included Seiji Ogawa, a researcher at Bell Labs who in 1990 had shown in rats that MRI could be sensitive to the amount of oxygen in the blood. He had referred to this as "blood oxygenation level dependent" (BOLD) contrast, a name that has since stuck as the description of the mechanism that is usually measured for fMRI. Although his studies in rats had not directly examined neuronal activity, he suggested in his 1990 paper that BOLD contrast could rival PET for the measurement of functional activity in the brain.[11] In order to test this idea in humans, Ogawa teamed up with Kâmil Uğurbil, a Turkish physicist who had worked with Ogawa previously at Bell Labs. Uğurbil had moved to the University of Minnesota and was building a research program focused on higher strength MRI systems; at the time, he and his group were developing a four tesla MRI scanner for humans, which would be delivered to Minnesota in 1990. By early 1991, the Minnesota group was attempting BOLD fMRI studies in humans, also using flashing goggles to stimulate the visual cortex; had their new MRI scanner not been damaged in transit from Germany, they would almost certainly have won the race to collect the first BOLD fMRI data.

The second group pushing to develop BOLD fMRI was from MGH, where Jack Belliveau had begun to work with another researcher at the center named Ken Kwong to develop an fMRI technique that did not require contrast injection. Kwong performed his first successful experimental run on May 9, 1991, as he recounts (see figure 2.5):

> The May 9, 1991 experiment was surprisingly smooth and trouble free. I borrowed the visual stimulator of a pair of flickering goggles from Dr. Belliveau's Gd-DTPA fMRI experiments, the same goggles on loan from Dr. Peter Fox who used them in stimulation rate experiments in the early 1980's (Fox and Raichle, 1985). I consulted Dr. David Kennedy on where the V1 region was located in the brain so he could help me pick the proper single brain slice for imaging. The stimulus design was a block paradigm alternating a baseline OFF epoch with an ON epoch for the flickering visual input for a total of 70 time points. ... It was a surprise that the visual cortex "lit up" with MR signal change at the very first run of the gradient echo experiment.[12]

Kwong attempted to submit his results for presentation at the annual meeting of the Society for Magnetic Resonance in Medicine (SMRM), which was the main conference attended by researchers in the field, but his submission was somehow lost in the mail. However, the director of the MGH-NMR Center, Tom Brady, was giving a major address at the meeting, and he included a mention of Kwong's results in his talk; August 12, 1991, was the first day that the world outside of Kwong's colleagues would learn of the new findings, and the first point at which the Minnesota researchers would realize how stiff their competition was. At the same meeting, Belliveau presented his work on fMRI using contrast agents (which would be published a few months later in *Science*), and he was also presented with the SMRM's Young Investigator award for his work.

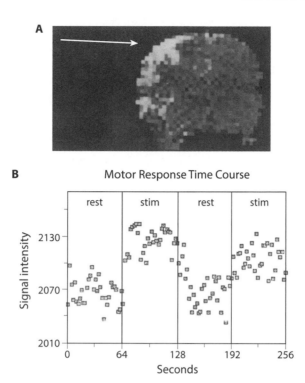

Figure 2.5. A figure from Kwong's landmark paper, showing that activity in the motor cortex (which controls hand movement) moved in concert with the subject making hand movements. Reprinted with permission of Kenneth Kwong.

The third group in the race was an unexpected dark horse: two graduate students from the Medical College of Wisconsin, a little known medical school in Milwaukee, Peter Bandettini and Eric Wong were working together to develop techniques to image brain activity, having been inspired by work from the French physicist Denis Le Behan that had suggested a method for identifying changes in blood flow by measuring the microscopic movement of water (known as *diffusion*) rather than the BOLD signal that Ogawa had discovered. When they saw Brady's talk at the SMRM meeting in August 1991, Bandettini says that "one might say that the light bulbs appeared above Eric's and my heads as we sat in the audience."[13] They also saw Belliveau present his work on fMRI activation using contrast agents at the same meeting. Once they got back to Milwaukee, they set to working on replicating the results from Kwong, and on September 14, 1991, they had their first successful experiment, with Bandettini as the volunteer.

In the wake of the 1991 SMRM meeting, the race was on to see who would publish their results first, since the first to publish often gets the lion's share of the credit for a scientific discovery. Kwong and his colleagues submitted their paper to the journal *Nature* in October 1991, but were surprised when the paper was rejected; the reviewers apparently didn't think that it went far enough beyond the earlier paper by Belliveau. Based on the comments from the reviewers at *Nature*, they revised the paper and resubmitted it to the *Proceedings of the National Academy of Sciences* (PNAS[14]), where it was published in June 1992. The Minnesota group had also sent their paper to *Nature*, only to have it rejected as well;[15] it was also submitted to PNAS, five days after Kwong's paper had been submitted, and was published in July 1992. The Milwaukee group, which had gotten the latest start on their BOLD fMRI experiments, actually ended up being the first to publish their paper. Instead of trying for a high-profile journal like *Nature*, they submitted their paper to the specialist journal *Magnetic Resonance in Medicine*, where it was published in June 1992, just edging out the Kwong PNAS paper. However, given that the MGH and Minnesota groups were working on BOLD fMRI well before Bandettini and Wong heard about their work from Brady's talk at the SMRM conference, credit for the discovery is generally given to those two groups even though their papers were published slightly later.

I was in graduate school in the early 1990s and had heard lots of hype about fMRI, but it wasn't available at the University of Illinois where I was a student. When I moved to Stanford as a postdoctoral fellow in 1995, I had not initially planned to do fMRI research, but I got pulled in by the excitement of this new technique. As we will see in the next chapter, fMRI would soon become the premier method used to ask questions about how the brain implements the mind.

CHAPTER 3

fMRI GROWS UP

Measurement is central to scientific research, and breakthroughs have often been driven by the development of new tools to measure the world. However, any new measurement tool has to be validated, to make sure that it is actually measuring what it claims to measure. In the case of fMRI, there was initial evidence that fMRI signals truly reflected brain activity, based on the fact that fMRI results aligned with what we already knew from neurological studies and from animal research: visual stimulation in different parts of the visual field causes activity in the appropriate areas in the visual cortex, motor actions lead to activity in the motor cortex, and so on. However, we still didn't have any direct evidence that linked fMRI signals to the activity of neurons, which meant that we were using the method without really knowing exactly what it was measuring. For this reason, many neuroscientists (particularly those whose research involved recording the activity of neurons directly in animals) were quite dismissive of fMRI   an attitude I experienced first-hand when I started interviewing for faculty jobs in 1998. Until we had evidence for a direct relationship between fMRI signals and the activity of neurons, the road for fMRI research would remain rocky.

Linking fMRI to Neurons

Soon after the discovery of fMRI, researchers started trying to understand the relationship between BOLD fMRI signals and the firing of neurons. If there is one thing that makes a system easier to understand scientifically, it is when the relationship between the inputs and outputs of the system is straightforward, which scientists call "linearity." A linear system is one in which the output of the system can be described by transforming and adding up the input. For example, my checking account is a linear system, in the sense that the balance is determined by adding up all of the individual transactions. In fact, it is a particular kind of linear system, that we call "time invariant," because it doesn't matter when the individual transactions occur—they still all get added together. Many systems in the world are linear— or at least, as scientists say, "linear to a first approximation," meaning that we can pretend they are linear and still do a pretty good job of explaining them, even if the model is not perfect. On the other hand, there are also many systems that simply can't be treated as linear. An avalanche occurs when too much snow falls on the side of a mountain, but you can't divide that snow into 10 parts and expect to get 10 small avalanches. In a linear system, more is more; in a nonlinear system, more is different.

In the years after the discovery of fMRI, researchers began to try to determine whether the fMRI signal was a linear function of the activity of neurons. That is, if the neurons fire twice as much, is the fMRI response twice as large? This question turns out to be quite challenging to answer because we can't directly measure how much the neurons are firing in a person—all we know is what kind of task we have provided the subject with, and how the fMRI signal changes. To study this question, Geoff Boynton and his colleagues at Stanford took advantage of the fact that neuroscientists have learned a lot about how neurons respond to different types of visual stimuli in the visual system of the monkey, and human brains work a lot like monkey brains. Boynton and his fellow postdoc Steve Engel each spent hours lying in the MRI scanner staring at checkerboard patterns as they flashed eight times a second, in order to measure how the response of their visual cortexes changed with the amount of time the checkerboard was presented for and the relative contrast of the black and white parts of the checkerboard—both factors known to affect the firing of neurons in the visual cortex in a well-characterized way. What they found was that the signal went up as the stimulus got longer and as the contrast was higher, exactly as they expected from the monkey research. The critical test of linearity was to see whether the response to a longer stimulus could be predicted just by adding together the responses to the shorter stimuli after shifting them in time, which is a critical prediction of the linear model. It worked—not perfectly, but well enough for most people to think that the linear model is a reasonable way to analyze fMRI data. The assumption of linearity is now fundamental to nearly all of the ways in which fMRI data are analyzed.

The work of Boynton and others provided a much tighter link between fMRI signals and the activity of neurons, but there was still a missing link: until someone recorded both single neuron activity and fMRI at the same time in the same brain, it would not be possible to say for sure that fMRI was reflecting the activity of single neurons. This challenge was taken up by Nikos Logothetis, a neuroscientist from the Max Planck Institute of Biological Cybernetics in Tübingen, Germany, who was an expert in studying how neurons respond in different parts of the monkey's visual system. He was becoming increasingly interested in more complex aspects of visual perception, such as our ability to pick out objects from cluttered backgrounds, and he realized that to understand these more complex phenomena, he needed to be able to study the entire system rather than just a few neurons, so when fMRI came around he started working on finding a way to do fMRI studies with monkeys while also recording the electrical activity of neurons. This was in some sense the "holy grail" of fMRI, since it combined the whole brain breadth of fMRI with the precision of single neuron activity recording. To appreciate just how difficult this is, keep in mind that electrical activity of neurons is recorded using electrodes connected to small wires, which register the electrical activity of the cell. Those changes are fairly small, on the order of microvolts (thousandths of a volt). It's also important to know that when metal is placed in an MRI scanner, the changes in the magnetic field used to create MRI images will cause current to start flowing through the metal, and that current can be much larger than the changes caused by neurons, making it almost impossible to see the tiny neuronal signals without some very sophisticated signal processing techniques. Logothetis and his team spent several years engineering a solution to these problems, and by 2000 they were able to successfully record from neurons in the brain of an anesthetized monkey while also performing fMRI.

The results from Logothetis's study provided direct evidence for a relationship between neural activity and fMRI,[1] and gave an "all clear" signal to many neuroscientists who had been leery of adopting the new technique without knowing precisely what it measured. The results also provided fMRI researchers like myself a way to answer the lingering questions from some of our colleagues about what fMRI actually measures. What Logothetis and his team did was to measure fMRI signals and neuronal activity simultaneously while the animal was presented with a flashing checkerboard. It may seem surprising that researchers could obtain useful data about brain activity from an anesthetized animal, but in fact the neurons in the visual cortex respond similarly between wakefulness and anesthesia. They saw clear evidence of activity in the visual cortex using fMRI, and this activity increased as the black/white contrast of the checkerboard was turned up (just as it had in Boynton's and Engel's brains). When they measured the firing of individual neurons they saw that it was also directly related to these changes in the BOLD signal; however, they found an even tighter link with something called the "local field potential" (LFP for short). The LFP is a measurement of changes in the electrical signal that happen more slowly than a neuron fires, and are thought to reflect the inputs to neurons rather than their firing. Logothetis's results have been built upon since 2001, most recently using fMRI in rats along with a technique called "optogenetics" that allows researchers to turn on specific kinds of neurons using light.[2] At this point, it is widely accepted that fMRI signals are a direct reflection of the activity of neurons, particularly of the inputs to neurons rather than their firing per se, even though there is still much to learn about exactly how it works.

Finding Modules in the Brain

Since the middle of the twentieth century, most neuroscientists have agreed that at least some psychological functions rely on specific areas in the brain. The advent of fMRI brought with it the promise of mapping out these localized functions with a remarkable degree of accuracy, and one of the first places this was used was in the study of how visual objects are recognized—specifically, faces. We already knew that the recognition of visual objects relies upon the bottom ("inferior") part of the temporal lobe, because damage to this region can leave people unable to recognize objects visually, even though they still have knowledge about the object and can identify it by touch. There was also some reason to believe that faces are processed differently from other objects, because there are rare cases of a syndrome called "prosopagnosia" in which the person is unable to recognize faces but is still able to recognize other kinds of objects. Early work using PET, particularly by the McGill University neuroscientist Justine Sergent,[3] had also shown that there were areas in the temporal lobe that were more involved in face processing than that of other types of objects.

Nancy Kanwisher is a neuroscientist at MIT who was captivated by the potential of fMRI to uncover the biology of the mind.[4] Using the same MRI scanner at the MGH-NMR Center on which Belliveau and Kwong had done their original fMRI studies, Kanwisher (along with her trainees Josh McDermott and Marvin Chun, who have both gone on to have impressive careers themselves) found a brain area that responded much more for faces compared with other types of stimuli.[5] The area was present in almost everyone they looked at (12 out of 15 people) in a location called the "fusiform gyrus" (see color plate 3), which runs along the bottom of the temporal lobe. This area didn't *only* respond to faces, but when they examined the response in this area to many different types of objects, in each case there was at least twice as much activity in the fusiform gyrus for faces compared with the other objects. They also showed that it was remarkably consistent over time—one of the participants (who was actually Nancy Kanwisher herself) was scanned multiple times over the course of six months, and in each case the pattern of activity was highly similar. They christened this area the "fusiform face area," or FFA for short. In the ensuing two decades, we have learned a lot more about the FFA, including the fact that there is not just one but several regions along the bottom of the temporal lobe that respond to faces (see color plate 3). We also now know that these areas are essential for face processing; this is the same area that was stimulated in the epileptic patient described in chapter 1 that so disrupted his ability to perceive faces. Kanwisher and others have also gone on to show that there are other parts of the temporal lobe that respond in a selective way for other types of stimuli, including body parts, words, and scenes.

Kanwisher and her colleagues made some very strong claims about the localization of face perception, which did not sit well with Isabel Gauthier, a vision researcher who had studied how our ability to recognize visual objects changes with practice. Her studies had trained people to identify a kind of artificial object called "greebles," which look somewhat like alien garden gnomes. Because greebles could be created using computer graphics to have many different kinds of visual features, they could be used to study how people become expert at distinguishing between them. What Gauthier found in her study was that people could get better at recognizing individual greebles with practice, and as this happened, the right fusiform area started to become activated by greebles just like it is for faces. This led her to pose a new theory: the FFA is not actually a "face area," but more like an "expertise area" (or, in keeping with the original acronym, a "flexible fusiform area"), becoming engaged whenever people recognize objects that they have lots of expertise with, especially when they have to distinguish different individuals within the category.

Gauthier set out to test this theory further by examining people who were highly expert at recognizing specific kinds of objects: bird watchers and car experts. There was already some reason to think that visual expertise involves different areas of the brain from those used in recognition of regular objects; in a couple of rare cases, a bird watcher lost the ability to recognize birds and a car aficionado lost the ability to recognize specific cars, without impairing their ability to recognize other kinds of objects. What Gauthier did in her study was to recruit a set of volunteers who were either bird watchers or self-proclaimed car experts, and then showed them a number of different types of objects, including faces, birds, and cars, along with other objects. The results confirmed the expertise hypothesis: the car experts showed a response to cars in the FFA, while the bird watchers showed a response to birds in that area; importantly, the response to the nonexpert objects was much lower. This seemed to conclusively show that the FFA was specialized for visual expertise, not for faces per se. However, Kanwisher and her collaborators were not convinced, raising a number of critiques of the data and their interpretation. Around the same time, however, another researcher came along whose work would both throw a new wrench into the FFA debate and light the path for a whole new way to think about fMRI data.

First Steps toward Decoding the Brain

Jim Haxby is in some ways the polar opposite of Nancy Kanwisher. Nancy is outwardly intense and energetic, while Jim's intensity lies beneath a soft-spoken and mellow surface. Haxby started his career at the National Institutes of Health, where he did research starting in the early 1990s using PET to study how different types of objects are processed in the brain; in fact, it was partly this work that inspired Kanwisher to look in the temporal lobe for face-related activity. When his group later began to use fMRI, he was captivated by the results from Kanwisher showing a seemingly very specific response to faces in the fusiform gyrus, but he did not believe their interpretation.[6] As his group began to do their own fMRI studies of face perception, he was struck by how these areas that were seemingly "specific" for faces actually showed substantial responses to other types of stimuli as well, bringing into question just how specific they were. Haxby and Kanwisher saw the same data, but came to strikingly different conclusions.

Haxby had an idea that would turn out to revolutionize how we analyze fMRI data. Until that point, fMRI researchers had looked only at "activation"—that is, how much more active a region was in one condition versus another. This information is usually presented in bright maps showing the regions where there was strong enough signal to be considered "statistically significant"— that is, where we are fairly sure that the difference in activity is not simply due to random fluctuations. Haxby had the idea to instead look at the entire pattern of activity across a region and ask whether it differed between conditions. As an analogy, think of a crowd responding to three candidates in a political debate by clapping (see figure 3.1). One could ask whether there is a section of the crowd that overall claps more for one candidate than the other two; this would be the equivalent of the activation analysis, finding the region of the crowd that is "selective" for one particular candidate. In this example, section 1 of the crowd claps much more strongly for the candidate denoted by the circle than for either of the other two; we would say that it was *selective* for the circle candidate. However, one can also flip this question on its head and ask: Can we tell which candidate is speaking, based on the pattern of clapping across the entire room? We refer to this as "decoding," and it has become a central part of fMRI analysis, as we will discuss throughout much of the rest of this book.

Understanding how decoding works requires a bit more detail about the nature of fMRI data. When we collect brain images using fMRI, those images are made up of a large number of three-dimensional cubes that we call "voxels" (think of pixels on a screen, but in three dimensions), each of which is 1–3 mm on each side. Each of these voxels contains millions of neurons, and the fMRI signal reflects an average of the activity across those neurons. If a voxel includes lots of neurons that fire when a face is present, then that voxel will show a high fMRI signal for faces, but it might also have a smaller number of neurons that prefer other types of objects, say houses or chairs. To push the crowd analogy a bit further, let's pretend that we are measuring the response in each section of the crowd using a microphone, which tells us on average how much clapping there is in that section of the crowd. In this example, each person represents a neuron, and each section of the crowd with a microphone represents a voxel. As in the bottom section of figure 3.1, we will probably see that there are some sections that clap much louder (on average)

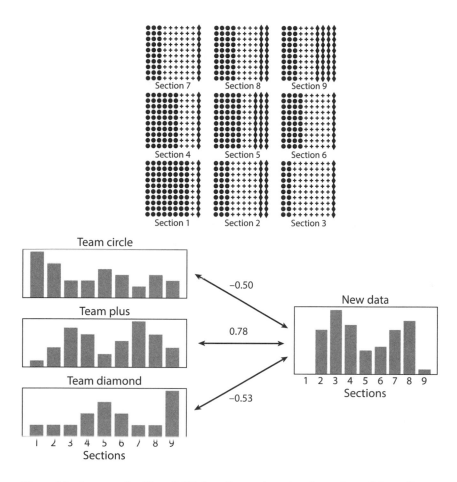

Figure 3.1. An example of how fMRI decoding works, using the analogy of the audience reaction to three different political candidates. The *top panel* depicts nine sections of the audience (which are analogous to voxels in fMRI); each section is made up of many individuals, each of whom claps for only one candidate (denoted by three different shapes: circle, plus, and diamond). These individuals represent the neurons within a voxel in fMRI. The *bottom left panel* shows a graph of the relative amount of clapping for each of the candidates across the nine sections; you can see that each candidate has a very different pattern of clapping across the sections. Now pretend that we have a new measurement of clapping (*bottom right*) and we want to decode which candidate is speaking. We can compute the correlations between the new pattern and each of the known patterns (which are the numbers presented next to the arrows); in this case, the new pattern is most highly correlated with the pattern that we observed when candidate plus was speaking, and thus we would predict that candidate plus was speaking when these data were collected.

for one candidate versus the others; that is the analog of higher activation for faces versus other types of stimuli. You can also imagine, however, that scattered across the crowd are people who have different preferences for the three candidates, and thus even in the areas that are not highly "selective" for one candidate, one might be able to see a difference in the pattern of clapping across those areas that would give a clue as to which candidate was speaking. The figure shows how, given a new measurement of the level of clapping across each section, we could decode who was speaking based on how similar the pattern across sections was to each of the known patterns.

Haxby used this idea to test his hypothesis that the processing of different types of visual information is "distributed" across the temporal lobe; that is, even though there are some areas that are highly responsive to faces versus other objects, those are not the only parts of the visual system that are processing information relevant to faces. In his study, he scanned volunteers while they looked at pictures of many different kinds of objects, including faces, houses, chairs, bottles, cats, shoes, and scissors.[7] Each person went through 10 different scans (which we call "runs") in which they saw each of the different types of object. To test whether he could decode the kind of object that a person was seeing, he first took the data from every other run (say, the odd runs) and measured the response in each voxel to each of the different types of object. Then, he took the data from the even runs, and for each one asked the following question: Given this pattern of activity, which one of the patterns is it most similar to from the odd runs? This allowed him to decode what kind of object the person was seeing. For example, if the current pattern (from one of the even runs) is most similar to the average pattern of activity for cats in the odd runs, then he would predict that the person was viewing a cat when those data were collected on the even run.[8] When Haxby applied this method to his data, he saw that he was able to decode what kind of object the volunteers were looking at, with an accuracy above 90%—in fact, for faces the accuracy was 100%! To test his claim that the processing of objects is distributed across the temporal lobe rather than localized, he looked at whether this decoding still worked even in the areas where activity was not selective for the specific object type; for example, could he distinguish between faces and other types of objects, even after he removed the voxels that responded more to faces than other object types? The answer was yes—he could still tell with very high accuracy when a person was looking at a face, even when he only looked within the voxels that were *not* part of the face area. As with most scientific debates, no single study provides a conclusive answer, and the controversy about the localization of face processing has continued since Haxby published his original paper. But, most importantly, his paper introduced the field to the idea of decoding mental content from fMRI data.

In chapter 1 I introduced the concept of "reverse inference"— the idea that one can tell what a person is thinking by looking at which brain areas are active—and explained how this kind of inference was problematic when it was applied in the *New York Times* op-ed about the 2008 election. As you have read this chapter, it may have occurred to you that the idea of reverse inference is really not very different from the concept of decoding that was seen in the work of Jim Haxby, and you would be correct: in each case we are using neuroimaging data to try to infer the mental state of an individual. The main difference is that the reverse inference that I ridiculed from the *New York Times* was based not on a formal statistical model but rather on the researcher's own judgment. However, it is possible to develop statistical models that can let us quantify exactly how well we can decode what a person is thinking about from fMRI data, which is the approach that Haxby and his colleagues took. Subsequent research has provided even more evidence of the power of fMRI to decode thoughts, which we will explore in much more detail in the next chapter.

From Modules to Networks

The debate over how faces are recognized was focused on whether the information was localized to one specific area or distributed across the temporal lobe, but there is an important point that this question obscures: the act of recognizing a face requires that these areas in the temporal lobe communicate with other parts of the brain that are involved in social processing, action, memory, and emotion, to name just a few relevant processes. As fMRI developed, researchers began to characterize how different parts of the brain communicate with one another.

Bharat Biswal arrived as a graduate student at the Medical College of Wisconsin in 1992, just after the team of Bandettini and Wong had performed their first fMRI scans.[9] For his project, he set out to understand the various sources of noise in the fMRI data, such as heartbeats and respiration, both of which can have substantial effects on fMRI signals. One of the tricks that he tried in order to better understand these signals was to take the time series of data from a voxel located in the left motor cortex, and then measure how the signals across the rest of the brain were correlated with this voxel during the scan. He expected that the voxels nearby the one he had selected would be correlated (since they should have neurons that react in a similar way), and indeed he saw this, but he also found something surprising: the motor cortex on the right side of the brain also showed signals that were highly correlated with the voxel in the left motor cortex (see color plate 4), even when the person was just resting in the MRI scanner and making no movements. In fact, the map that he obtained by measuring the correlation between the left and right motor cortexes during resting fMRI looked very much like the map that he obtained when he compared moving both hands to resting, meaning that the motor cortex could be identified using resting-state fMRI even when the person was not moving at all. Biswal published these results in 1995, but it took about a decade for researchers to realize just how important his ideas were. As we will see in chapter 5, the study of the brain in people simply lying in an MRI scanner at rest is now one of the most powerful techniques in human neuroscience.

As researchers began to study how different parts of the brain are connected to one another, they took advantage of several new techniques. When Biswal measured correlations in fMRI signals across different brain regions, he was measuring what we call "functional connectivity"—the degree to which activity in different brain areas moves together over time. This does not tell us whether the regions are "structurally connected" by the white matter of the brain, which is the cabling that connects different brain areas. It could be that two areas are connected to one another directly by white matter (which we call a "tract")—like Highway 101 that directly connects Los Angeles and San Francisco, but the functional connection could instead go through other areas in more than one step—like driving to San Francisco from Los Angeles via Las Vegas. If we want to understand the wiring diagram of the brain, then this is a crucial question. Historically, the tracing of tracts in the white matter has been studied in animals by injecting a radioactive "tracer" in one region and then looking at where it ends up in the rest of the brain as it moves along the brain's axons. We can't do this in living humans, but once again MRI has come to our rescue, through a technique called "diffusion weighted MRI" (or DWI for short) that images the movement of water molecules in the brain. We can use DWI to image white matter tracts because the axons that make up the tracts are covered with a fatty material (called *myelin*) that insulates them, like the plastic sheath that covers an electrical wire. Because it's difficult for water to pass through the cell membranes and their fatty myelin insulation, it tends to move more easily in the direction of the axon rather than across it. By measuring the diffusion of water in many different directions, we can infer the white matter connections between regions in the brain, using a technique that we call *tractography*.

By putting together information from functional connectivity measurements via fMRI and structural connectivity measurements using DWI, we can start to trace out what we call the "connectome": the catalog of connections between all of the different areas of the brain. Many people are familiar with this term through Sebastian Seung's book by the same name, and his TED talk titled "I Am My Connectome," by which he meant that everything that makes each of us unique is stored in the specific connections between neurons in our brain. Seung's work focuses on specific connections between individual neurons, what we might call the *microscopic* connectome, which can only currently be studied in nonhuman animals. In neuroimaging we instead focus on the large-scale connections between different areas in the brain, or what we call the *macroscopic* connectome. In the end we hope that these two lines of research will converge, though it will always be difficult (if not impossible) to study the microscopic connectome in humans. The importance of understanding the brain's wiring diagram led the National Institutes of Health (the major funding agency for biomedical research in the United States) to spend US$30 million from 2010 to 2014 on the Human Connectome Project, with the goal of providing a detailed map of human brain connectivity. Over the course of that period, the Human Connectome Project collected MRI data, psychological testing, and genetic material from 1,200 individuals, and the data were made openly available to scientists around the world (just as they had been for the Human Genome Project). These data have led to several important breakthroughs regarding brain function, as well as to a new atlas for the human brain that has discovered new areas and characterized the differences between people in how their brains are organized.

The increased interest in connectomics has coincided with the broader development of what has come to be called "network science"—the science of complex networks, which can range from connectivity in the brain to friendships on Facebook to flights between airports.[10] Many people will be familiar with the idea of "six degrees of separation," made particularly well known by the demonstration that nearly every actor in the Internet Movie Database can be linked to Kevin Bacon by six or fewer costars.[11] What this phenomenon highlights is that complex networks often have a particular kind of structure that makes communication across the network very efficient; these are then known as "small-world" networks. A small-world network is one in which there are a small number of elements (which could be people, brain areas, or airports) that are highly connected, which we refer to as "hubs." For example, Heathrow and Newark International airports are both hubs in the sense that they have flights to many different airports (including other hubs), whereas the airport in Ithaca, New York, or Fresno, California, may only have flights to one or two different airports. Neuroimaging research has shown that the human brain has many of the features of a small-world network, and the tools of network analysis have been used to make a number of interesting new discoveries about brain function, which we will delve into more deeply when we discuss brain connectivity in chapter 5.

Growing Pains

Whenever a new measurement tool emerges, the scientific community often struggles to understand how to work with the data and what their limits are, and fMRI has been no exception. In fact, the high profile of fMRI research has made it an attractive target for researchers looking to criticize it.

One of the major challenges for the analysis of fMRI is the fact that we collect so many measurements at once. In comparison to a psychology study where we might just measure 5 to 10 different variables, in fMRI we regularly collect data from more than 100,000 locations in the brain. The unique challenges of dealing with such big data were highlighted in one of the most amusing episodes in the history of fMRI.

In 2009 I was a member of the program committee for the Organization for Human Brain Mapping, which is responsible for vetting submissions to make sure that they meet the standards of the organization before accepting them for presentation at the annual meeting. One of the criteria for rejecting a submission is if it is a joke, and one particular submission was flagged for this reason by one of its reviewers. The title of the abstract was "Neural Correlates of Interspecies Perspective Taking in the Post Mortem Atlantic Salmon. An Argument for Multiple Comparisons Correction," which certainly doesn't sound like a very funny joke, but a closer reading of the submission showed why the reviewers had been concerned:

> **Subject**. One mature Atlantic Salmon (Salmo salar) participated in the fMRI study. The salmon was approximately 18 inches long, weighed 3.8 lbs, and was not alive at the time of scanning.

> **Task**. The task administered to the salmon involved completing an open-ended mentalizing task. The salmon was shown

a series of photographs depicting human individuals in social situations with a specified emotional valence. The salmon was asked to determine what emotion the individual in the photo must have been experiencing.[12]

What the researchers, Craig Bennett and his colleagues, had done was to put a dead salmon in an MRI scanner, present it with a "task," and record fMRI data. They then analyzed the data in a particular way, and found that there was apparently activation in the salmon's brain in response to the task (see color plate 5). The authors did not do this to demonstrate some kind of after-life mental capacity in the salmon; rather, they did it to prove a critical point about analyzing fMRI data—one which had been known for many years, but had nonetheless been neglected by many researchers in the field.

Remember that fMRI data consist of measurements from many small cubes ("voxels") across the brain. In a standard fMRI scan we would collect data from anywhere between 50,000 and 200,000 voxels. In order to determine which parts of the brain respond to our task, we compute a statistic at each voxel, which quantifies how much evidence there is that the voxel's signal fluctuates in the way we would expect if it were actually responding to the task. We then have to determine which regions show a strong enough response that they cannot be explained by random variability, which we do using a statistical test. If the response in a voxel is strong enough that we don't think it can be explained by chance, then we call it a "statistically significant" response. In order to determine this, we need to determine how willing we are to accept false positive results—that is, results that are called statistically significant even though there is no actual signal in the data (known technically as "type I errors"). There is also another kind of error that we can make, in which we fail to find a statistically significant result even when there is a true effect in the voxel; we call this a "false negative" or "type II error." These two types of statistical errors exist in a delicate balance—holding all else equal, increasing our tolerance for false positives will decrease the rate of false negatives, and vice versa.

The usual rate of false positives that we are willing to accept is five percent. If we use this threshold, then we will make a false positive error on five percent of tests that we perform. If we are doing just a single test, then that seems reasonable—19 out of every 20 times we should get it right. But what if we are doing thousands of statistical tests at once, as we do when we analyze fMRI data? If we just use the standard five percent cutoff for each test, then the number of errors that we expect to make is 0.05 multiplied by the number of tests, which means that across 100,000 voxels we are almost certain to make thousands of false positive errors, and this is in fact what Bennett and his colleagues found. They wrote: "Can we conclude from this data that the salmon is engaging in the perspective-taking task? Certainly not. What we can determine is that random noise in the [fMRI] timeseries may yield spurious results if multiple comparisons are not controlled for." Unfortunately, that part of their conclusion was often lost when the results were discussed in the media, resulting in a misleading impression that fMRI data were untrustworthy.

In fact, neuroimaging researchers have understood this problem of "multiple comparisons" since the days of PET imaging, and statisticians have developed many different ways to deal with it. The simplest (named after the mathematician Carlo Bonferroni) is to divide the rate of false positives for each test (known as *alpha*) by the number of tests. This controls the false positive rate, but is often overly conservative, meaning that the actual rate of false positives will be less than the five percent rate. This is problematic because, as I mentioned before, there is a seesaw relation between false positive and false negative rates, so an overly conservative test will also cause a high number of false negative errors, meaning that researchers will fail to find true effects even when they are present. However, there are a number of methods that have been developed that allow researchers to control the level of false positives without being overly conservative. While it was common to see fMRI papers published without appropriate statistical corrections in the early days of imaging, today nearly every paper reporting fMRI results will use a method to correct for multiple comparisons.

Is fMRI Just Voodoo?

Another well-publicized critique of fMRI research centered on the study of how differences between people in their behavior relate to differences in their brain activity, which is commonly used in neuroimaging research. An example of this can be found in a study that my colleagues Sabrina Tom, Craig Fox, and Chris Trepel and I did, which was meant to understand why some people are more willing to take risks than other people;[13] I will discuss this study again in chapter 7. To examine this, we presented the 16 subjects with a number of different gambles (such as a 50/50 chance to win $26 or lose $14) while they were scanned with fMRI, and asked them whether they would take that gamble. To make sure that they treated it like a real decision, after the scan was finished we randomly selected a few of the trials and then, if they had said "yes" to the gamble, we flipped a coin to play that gamble for real money. On average, people are averse to losing, which means that most people won't say "yes" unless the amount they could win is about twice the amount they could lose. However, we also found that there was a great amount of variability across people in their degree of loss aversion: some people would agree to accept gambles where the amount they could win was just barely above the amount they could lose (say a 50/50 chance to either win $14 versus lose $12), whereas other people required the amount to be won to be several times as large as the amount to be lost before they would agree to the gamble.

We set out to understand this by analyzing how their brains responded to increasing gains and increasing losses, and we found that there were some regions in the brain where there was a very close relation between the loss aversion that we saw in their choices, and what we called "neural loss aversion," reflected by the degree that the brain was more turned on by gains than it was turned off by losses. In fact, we saw strikingly strong correlations between behavior and brain activity. The relation between the brain and behavior is defined using a statistic called a "correlation coefficient," which goes from one (meaning that the variables are perfectly related), to zero (meaning that they have no relation), to negative one (meaning that they move

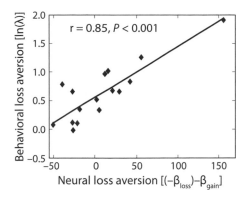

Figure 3.2. Figure from our 2007 paper, showing a puzzlingly high correlation between behavioral and neural loss aversion. The high strength of this correlation was due in part to the nonindependence of the analysis. From "The Neural Basis of Loss Aversion in Decision-Making Under Risk" by Sabrina M. Tom, Craig R. Fox, Christopher Trepel, Russell A. Poldrack, *Science*, 26 January 2007: 515–518. Copyright © 2007 by AAAS. Reprinted with permission from AAAS.

exactly in the opposite direction). We found that there was a correlation of 0.85 between the brain signals and the subjects' behavior (figure 3.2 reproduces a figure from that paper, showing this high correlation). This should have seemed too good to be true—but as the physicist Richard Feynman once said, "The first principle [of science] is that you must not fool yourself and you are the easiest person to fool."[14] What we should have done was to return the study and replicate the finding before publishing it, but running an fMRI study costs many thousands of dollars and could take several months. We let the excitement of discovery overcome our skepticism, and submitted the paper to the high-profile journal *Science*, where it was published in 2007.

Around that same time, Ed Vul and Hal Pashler were working on a paper that would rattle the fMRI world and bring about a serious public crisis. Vul and Pashler had seen other researchers present data just like those from figure 3.2, and it had left them thinking that it must be too good to be true. Their skepticism was driven by the fact that true correlations of that magnitude are only possible if the underlying variables being correlated are highly reliable—where "reliable" means that if we measure them twice, we should get the same number. In fact, it is a statistical rule that the true correlation between two variables can't be much higher than the reliability of either measurement. We know that the reliability of psychological measurements like our gambling task is rarely above 0.8, so that should have already given us pause. We and others had also done research on the reliability of fMRI measurements across time, and we knew that it also rarely exceeds 0.8 and is often much lower. So how did studies find such high correlations?

Vul and his colleagues intuited that this was because of a statistical error known as "nonindependence" or "circularity." Imagine that I were to tell you that I had made a novel discovery that Stanford students had much higher SAT scores than the general population. You would immediately laugh at me and say "of course they do!" The average SAT score of Stanford students is necessarily higher than that of the general population, because the SAT score is one of the variables that goes into selecting them for admission. In the case of fMRI, we perform a large number of correlation tests across all of the voxels. If we then take the strongest ones and plot them, they will seem impressively large, but that's because we have guaranteed it to be the case. In fact, one can find impressively large correlations from completely random data using this kind of analysis (as we showed in a paper published in 2017).[15] One can imagine a different way of doing the analysis, in which we use one set of data to find the area of interest, and another separate set of data to compute the correlation, which Vul called an "independent" analysis; that approach does not suffer from the circularity that is present in the nonindependent approach.

Vul focused on the domain of social neuroscience, in part because brain-behavior correlation analysis was very common in that area. He requested information from a large number of authors about how they had done their analyses, classifying each analysis as either independent or nonindependent. When he combined the results across studies, he found that the correlations reported in studies using nonindependent analyses were much higher than those using independent analyses, and nearly all of the studies with correlations above 0.8 had used nonindependent analyses. In an aggressive challenge to the field,

Vul and colleagues titled their paper "Voodoo Correlations in Social Neuroscience."[16] I received a copy of their paper some time in 2008, before it had been published. As I read it, I had a sinking feeling; even though our 2007 *Science* paper was not one of the ones included in their list, I could see that we had made exactly the same error of nonindependence. I still thought that the bulk of our results were solid, since the correlation presented in our figure was mostly for illustration, and the underlying results had survived stringent correction for multiple comparisons, but I was still left with unease about the fact that our figure had misled readers. I set out to reanalyze our data in a way that did not suffer from nonindependence, using a technique called "cross-validation" (that we will discuss in more detail in Chapter 4) which allows us to test how well we can make predictions to independent data. What I found was that the correlations still held when using a proper independent analysis, but were smaller by about 40% compared with the nonindependent analyses.[17] Thus, our conclusions still held, but they were less impressive than we originally thought.

It is rare that a debate about data analysis methods in science reaches the pages of *Newsweek*, but that's indicative of the intense firestorm that this paper started. It also spurred a set of spirited published responses from fMRI researchers and statisticians, most of whom agreed with the substance of the Vul critique, if not the alarmist tone. But not all of them. Matt Lieberman (who at the time was a faculty colleague of mine at UCLA) could fairly be considered to be the primary target of the Vul article. It was his 2003 paper that was the first one listed as an example of supposed "voodoo" in Vul's paper, and Vul also claimed to have discovered a statistical error in one of the analyses reported by Lieberman's group. Lieberman and his colleagues Elliot Berkman and Tor Wager responded, stating that "Much of the article's prepublication impact was due to its aggressive tone, which is nearly unprecedented in the scientific literature and made it easy for the article to spread virally in the news."[18] They went on to try to argue that researchers didn't really mean to overstate the strength of their correlations—but this is a difficult argument to make when, as Vul pointed out in his counter-response,[19] some researchers had referred to their correlations in press releases as "insanely strong."

The critique from Vul and colleagues shook me to the core, and drove me to rethink how we did fMRI research. I had always thought of myself as a reasonably savvy and careful researcher, but the fact that I had let myself be fooled by those seemingly strong correlations showed that I still had a long way to go. In the ensuing years we have become much more attentive to the ways that our data analysis methods can lead us astray, and we think that these improvements have increased the reliability of our research.

CHAPTER 4

CAN fMRI READ MINDS?

One of the films that I remember most vividly from my teenage years is *Brainstorm*, in which Christopher Walken plays a scientist who has developed a device that can record the entirety of one's conscious experience and allow others to replay it. When we think of the concept of "mind reading" it is often in the context of this sort of science fiction, but to listen to some fMRI researchers, it is very close to being science fact. In 2009 the television journalist Lesley Stahl interviewed Marcel Just of Carnegie Mellon University, an early researcher in the field of fMRI decoding:

> STAHL: Do you think one day, who knows how far into the future, there will be a machine that will be able to read very complex thoughts, like "I hate so-and-so," or "I love the ballet because . . ."?
>
> JUST: Definitely, and not in twenty years, I think in three, five years.
>
> STAHL (somewhat incredulous): In three years?
>
> JUST: Well, five [laughs].[1]

Fortunately or unfortunately, depending on your point of view, we are not there yet—but research has nonetheless started to uncover findings that verge on what many people would consider legitimate examples of mind reading.

Deciphering the Language of the Mind

Around the time that I published my critique of reverse inference in 2006, many researchers in our field were becoming very interested in seeing just how far we could push the limits of fMRI in order to determine what a person is thinking about— which researchers sometimes audaciously call "mind reading" but is more accurately termed decoding. As I presented in the discussion of Haxby's work on face decoding in the previous chapter, the goal is similar in spirit to the kind of reverse inference used in the "This Is Your Brain on Politics" piece that I mentioned in chapter 1—to determine the contents of a person's mind from brain activity. However, the approach is very different because it uses statistical tools to quantify exactly how well we can actually decode what a person is thinking about or experiencing.

One way to think about brain decoding is that it is trying to translate between two languages: The natural language of humans, and the biological "language" of thought in the brain. This translation happens indirectly through a set of sensors (such as an MRI machine), since we can't directly "hear" the brain speak its language. As we will discuss later, this kind of translation is likely to be very difficult, if not impossible, using fMRI alone. However, a potentially more achievable goal is to develop a dictionary that maps between patterns of fMRI signal and particular mental states or experiences. This is how Jack Gallant, one of the leaders in the field of fMRI, thinks of the problem:

> In principle, you can decode any kind of thought that is occurring in the brain at any point in time. ... you can think about this like writing a dictionary. If you were, say, an anthropologist, and you went to a new island where people spoke a language that you had never heard before, you might slowly create a dictionary by pointing at a tree and saying the word "tree," and then the person in the other language would say what that tree was in their language, and over time you could build up a sort of a dictionary to translate between your language and this other foreign language. And we essentially play the same game in neuroscience.[2]

Such a dictionary wouldn't give us back complete sentences in the brain's language, but it would at least give us the words—and that's often enough to get one pretty far. Nearly all of the work done to date in the development of fMRI mind reading can be thought of as working toward such a dictionary, and at this point I would say that our dictionary for simple, common words in the brain's language is pretty good.

A Penny for Your Thoughts

The research by Haxby and his colleagues showed that one can decode the contents of visual perception from fMRI signals with very high accuracy. However, this was not particularly surprising to many researchers, since we already knew that visual objects are processed in the temporal lobe, and that the responses of neurons in these areas to visual objects occur even in animals that are under anesthesia—meaning they don't even require the animal to be conscious of the objects. What about conscious thoughts? This question was taken up by John Dylan Haynes, who in the mid-2000s was a postdoctoral fellow at the Wellcome Trust Centre for Neuroimaging in London, which is known to researchers in the field as "the FIL" (for its former name, the Functional Imaging Laboratory). The FIL was (and remains) one of the top neuroimaging centers in the world, and at the time Haynes was working with a young professor named Geraint Rees who had already made a name for himself studying how we are conscious of visual objects. Together they published a set of studies that showed how neuroimaging could be used to decode the contents of our conscious visual experience. In one of these, they presented volunteers with visual stimuli of a different color to each eye,[3] which results in something called "binocular rivalry" in which the person's conscious perception switches occasionally between the two eyes. They recorded what color the person experienced at every point in time during the fMRI scan. The results showed that they could decode from fMRI data which of the colors the person was experiencing at each point in time with relatively high accuracy.

Many people might not consider the decoding of visual experience to qualify as full-blown "mind reading," but in Haynes's next study it would become much harder to dispute that label. He wanted to determine whether he could decode a person's intentions about a future action from fMRI data. To do this, he gave people a task where they had to choose whether to add or subtract pairs of numbers. In all the trials in the experiment, people first saw a cue that told them to decide whether to add or subtract the numbers on that trial, and then a few seconds later the numbers appeared on the screen. They were given some time to do the addition or subtraction, and then shown a set of probe numbers that included the results of both addition and subtraction, and were told to choose the result from their chosen arithmetic operation. Haynes then asked whether he could use the fMRI signal from the initial cue period (when people were simply thinking about what they were going to do) to predict which of the operations each person would actually choose (which Haynes knew based on which number the person chose on the probe). The results were striking: There were several places in the brain where brain activity was predictive of what action the person would make in the future. The prediction was not perfect—it was around 70% accurate, where random guessing would result in 50% accuracy—but nonetheless it provided a powerful example of how fMRI could decode even very private abstract thoughts.

The work by Haxby, Haynes, and others had an important limitation: in each case, the prediction was person specific. That is, the prediction was performed by taking some data from a person and using it to train a statistical model that could then make predictions based on other data from the same person. There was no testing of the ability to generalize from one person to another. In this way, there was a serious mismatch between what the studies actually showed and some of the discussion in the press, which raised concerns about the use of fMRI to predict crimes or other behaviors. After seeing Haynes's results, I became interested in asking the question of whether it was possible to decode mental states from fMRI data, even when we had never seen the particular person's brain. To ask this question, we took data from 130 people who had each participated in one of eight different studies in my laboratory. The studies involved different cognitive tasks ranging from reading words to choosing monetary gambles to learning new categories of objects, and it occurred to me that we could potentially train a statistical model to predict which of these tasks they were doing. To do this I teamed up with Steve Hanson and Yarick Halchenko from Rutgers University, who are experts in developing these kinds of models. The field that they work in goes by a number of names including "machine learning," "statistical learning," and "pattern classification"—but you can think of it as the science of how to make good predictions from data. In this case, we want to use brain imaging to predict what a person is thinking, but it's the same set of statistical tools that Facebook uses to recognize faces in photos and Google uses to predict which e-mails are spam and which are not.

What we did was to use a method that is standard in machine learning, known as *cross validation*, which had also been used by Haxby and Haynes before us. The goal of cross validation is to let us tell how well our statistical models can generalize to new data. In principle one could test this by collecting another data set and seeing how well the model from the first data set can generalize to the second, but often we just can't collect another data set. The idea behind cross validation is simple but very powerful. First, we break the data into subsets—for simplicity, let's say that the subsets are as small as possible, which would be individuals in the data set. Thus, for the 130 subjects in our study, we would have 130 subsets. Then, we train the statistical model separately, using all of the data *except* those from one left-out subset, and then test the model on the left-out data. For example, on one round we would fit the model on subjects 1–129, and then test it on subject 130, while in another we would fit the model to all subjects except for subject 129 and then test on subject 129, and so on for all possible subsets. This particular technique is called "leave-one-out" cross validation, for obvious reasons. For each left-out data set, we test how well our predictions match with reality. In this case, we know which of the eight possible tasks each person was performing, and the statistical model gives us a prediction for which task the person is doing based on his or her brain activity; we count up how often the prediction is correct, and that is our measure of accuracy. We were able to predict the task for the left-out subject with about 80% accuracy—for comparison, if we were just guessing we would expect to get it right about 13% of the time. This study was the first to demonstrate conclusively that it is possible to decode mental states from fMRI data, even when our statistical model has been trained on other people, laying the groundwork for later research that would push brain decoding even further.

Reading the Mind's Eye

The research on fMRI decoding that we have discussed so far has focused on the ability to choose between a small number of possible states—for example, eight different types of images in Haxby's study (chapter 3)—but true mind reading implies the ability to reconstruct any arbitrary thought or image from brain activity. Two studies published in 2008 provided a hint that this might be possible, building on earlier work by a French group led by Bertrand Thirion. The approach that these two studies used was different from the previous work in a particular way: while the previous decoding studies had used generic machine-learning methods (which could have just as easily been used to predict what you are planning to buy on Amazon.com), these newer studies used models that were specifically built to mimic the structure of the human visual system.

Kendrick Kay, working with Jack Gallant at Berkeley, wanted to test whether it was possible to identify natural images amongst a large number of possibilities. Kay and his colleague Thomas Naselaris each spent several hours in an MRI scanner, ultimately viewing almost 2,000 different natural images.[4] They used the brain responses on 1,750 of these images to create a statistical model that was informed by the structure of the human visual cortex. This "quantitative receptive field model" basically tried to figure out, for each voxel in the visual cortex, what parts of the visual world it was sensitive to, which we generally refer to as a "receptive field" (figure 4.1)—you can think of this as a map of which parts of the visual world a particular part of the brain is paying attention to. By combining these across many voxels, they were able to generate a model of the entire visual cortex. Then, using the data from the remaining 120 images, they asked

Subject S1, voxel 21672, area V3

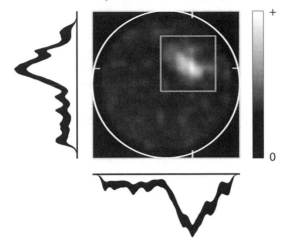

Figure 4.1. A reconstruction of the "receptive field" for a single voxel in Kay's 2008 study. The plot illustrates how different voxels in the brain respond to stimulation in different parts of the visual field—each voxel has its own small area. The bright spot shows the part of the visual scene that this one voxel was sensitive to—in this case, a patch slightly right of the center of the visual field. Unpublished image courtesy of Kendrick Kay.

whether the model could identify which image was being viewed (out of the entire set of 120) using only the fMRI data. To do this, they took the actual brain activity for each image and compared it with the predicted brain activity from the model for each of the 120 images, asking whether the actual brain activity was closest to the predicted activity for the actual image being viewed versus the other 119 possible images. If one were guessing, one would get it right only less than 1% of the time, but for both of the subjects Kay and Gallant were able to choose the correct picture with high accuracy (92% for one subject, 72% for the other). This showed that the ability to decode visual image content went far beyond the small number of categories studied in the previous work, but it didn't quite get to full-blown reconstruction of an arbitrary image.

A second study by Yukiyasu Kamitani and his colleagues in Kyoto, Japan, moved a step closer to reconstructing a viewed image.[5] They used an approach similar in spirit to the one used by Kay and Gallant, in which they built a large set of simple

Figure 4.2. An illustration of image reconstruction using fMRI. The *left panel* shows a figure adapted from the 2008 paper by Miyawaki and colleagues. The *top row* shows the actual images that were presented to the subject, and the *bottom row* shows the visual pattern that was reconstructed on the basis of the subject's brain activity. The *right panel* shows images adapted from the 2009 paper by Naselaris and colleagues, with actual images on top and the closest matching actual image on the bottom. The numbers in the bottom-right corners of the lower images are a measure of the accuracy with which the model could predict brain activity. Left panel reprinted from *Neuron* 60, no. 15, Yoichi Miyawaki, Hajime Uchida, Okito Yamashita, Masaaki Sato, Yusuke Morito, Hiroki C. Tanabe, Norihiro Sadato, and Yukiyasu Kamitani, "Visual Image Reconstruction from Human Brain Activity Using a Combination of Multiscale Local Image Decoders," 915–29, Copyright 2008, with permission from Elsevier. Right panel reprinted from *Neuron* 63, no. 6, Thomas Naselaris, Ryan J. Prenger, Kendrick N. Kay, Michael Oliver, Jack L. Gallant, "Bayesian Reconstruction of Natural Images from Human Brain Activity," 902–15, Copyright 2009, with permission from Elsevier.

decoders, each of which learned to decode the signal in a small part of the visual image based on data from a small portion of the visual cortex. They then used a machine-learning method to learn how to combine these in order to best reconstruct the image that had been presented. The results obtained on simple geometric shapes were impressive (figure 4.2), including a reconstruction of the word "neuron," which was the name of the journal where the work was ultimately published.

The final step toward full reconstruction of a natural image also came from Gallant's group at Berkeley, this time led by Thomas Naselaris.[6] This study made two major advances that allowed these researchers to reconstruct entire natural images. First, they used the idea of Bayesian analysis, in which the data are combined with prior knowledge to determine the best possible reconstruction of the data. In this case, the prior knowledge consisted of six million images selected at random from the Internet. Essentially what they did was use the fMRI data to create a predicted image, and then ask which of the six million actual images was most similar to that predicted image. The second advance was to include information about the semantic category of the scene (such as animate vs. inanimate or indoor vs. outdoor), which was generated by hand for each of the training images. They then used data from a higher visual area, which is known to respond to object catgories, to detect which category was present. Putting these together they were able to acheive very impressive "reconstructions" (see figure 4.2)—each of which was actually a selection of the closest image from among the six million images. This approach was later used by the same group to reconstruct movies as well. While one can question whether this really counts as "reconstruction," it's likely that any model that is going to successfully decode images from brain activity will have to take advantage of strong prior information about the features of natural images, just as our visual system takes advantage of such information when we recognize images.

Decoding Consciousness in Brain Injury

Of all of the possible things that a neuroscientist fears, a severe traumatic brain injury has to be at the top of the list. A blow to the head from a car accident or a fall can send a perfectly healthy and cognitively intact person into a mental no-man's-land in a split second, and the best one can hope for after such an injury is often basic self-sufficiency; a complete recovery of intellect and personality after a severe brain injury is often out of the question. Advances in emergency treatment have allowed many more people to survive serious brain injuries, but these survivors are usually left in a state of altered consciousness for an extended period of time. The lowest level of consciousness is *coma*, in which people are completely unresponsive (even to painful stimulation) and do not open their eyes. Coma is not the same as "brain death," which represents an even deeper and irreversible level of damage to the brain; the brain of a comatose person still has electrical activity, though it is highly abnormal. A person in a coma will usually over time begin to show some signs of brain function, such as opening the eyes, but often remain nonresponsive and doesn't show any outward signs of conscious awareness; this is referred to as a *vegetative state* and can sometimes last for years if the person is fed via a tube and otherwise cared for. In other cases, the person begins to show increasing signs of consciousness, and can often exist in what is now called a *minimally conscious state*—in which he or she drifts in and out of lucidity, sometimes being able to interact with others while at other points being unresponsive.

Because people in a vegetative state appear to be unconscious and show abnormal electrical activity in the brain when measured using electroencephalography (EEG), it was long assumed that they did not have conscious awareness. However, we know of cases where a person can be fully conscious yet seem completely unresponsive—this occurs in "locked-in syndrome," a very rare syndrome caused by damage to the brainstem, which leaves patients fully conscious but unable to make any movements other than blinking and moving their eyes. What if some of the people thought to be in a vegetative state were actually aware but unable to express themselves, like the locked-in patients? Adrian Owen has spent the past two decades trying to answer this question. He is a cognitive neuroscientist who started out studying basic cognitive processes using PET imaging, but at some point became obsessed with understanding conscious awareness and using fMRI to identify it in people suffering from disorders of consciousness.

Owen's solution was surprisingly simple. He placed individuals in the MRI scanner and then asked them to do one of two different things: either imagine playing tennis, or imagine walking through their house. He chose these two different tasks because he knew from studies of healthy people that the two tasks should evoke very different patterns of activity if the person performed the task properly. They tested the method on a 23-year-old woman who had suffered a severe brain injury in a car accident five months earlier and who remained in a vegetative state, completely unresponsive to stimulation. While her unresponsiveness suggested that she did not have intact cognitive function, fMRI showed a different story. When she was told to imagine playing tennis there was activity in her premotor cortex, while there was activity throughout the network that is engaged during spatial navigation in healthy people when she was told to imagine navigating her house (see color plate 6).[7] This landmark finding inspired a much broader analysis, which has shown that the proportion of people in a vegetative state who pass the test for conscious awareness is relatively low—in a subsequent study led by Martin Monti, 5 out of 54 individuals in a vegetative state showed evidence of awareness.[8]

The use of neuroimaging to detect conscious awareness in people with brain injuries is a major advance that shows the real world utility of fMRI decoding. At the same time, these studies have raised some difficult ethical and medical questions. Foremost, the ability of people in a vegetative state to answer questions means that they could in principle be asked the most difficult possible question of all: Do they wish to continue living? These individuals are not able to feed themselves and thus require feeding through a tube, which they could in theory ask to be stopped. How would we decide whether they have sufficient reasoning ability to make this judgment, and how would we respond? It's worth noting that while healthy people would assume that many individuals in such a state would choose to end their own lives, there is some evidence to suggest the contrary. In particular, a study of individuals with locked-in syndrome that assessed their subjective well-being showed that the large majority claimed to be happy with their life, and only 7% expressed the desire to be allowed to die.[9] Caregivers and physicians also need to think deeply about how the knowledge of a person's state of conscious awareness and the ability to ask him or her questions using fMRI would change the way that they treat the person. Would they try to seek consent for a risky medical procedure that has the potential for negative outcomes? We also need to know whether, and how well, these markers of consciousness predict recovery in the future, and to decide whether knowing about a person's cognitive status is actually useful if it does not provide any useful clinical guidance.

Are You Really in Pain?

I once attended a lecture by an eminent pain researcher, who started his talk by making what may seem like an obvious point: pain is in the brain. What he meant was that the aversive nature of pain depends on how our brain responds to the input from our peripheral nerves that carry impulses to the brain. When I accidentally smash my thumb with a hammer, specialized nerve receptors in my thumb send a message to my brain telling it that something bad has happened, and I experience those impulses as the aversive experience of pain. Pain is annoying but essential, as it alerts us to the need to protect the injured area to prevent further injury, and to seek treatment if needed. The utter importance of pain is evident in the plight of people who suffer from *congenital analgesia*—that is, they are born without the ability to feel pain. Mo Costandi described the case of Ashley, a teenager born with this disorder:

> As a newborn, she barely made a sound, and when her milk teeth started coming out, she nearly chewed off part of her tongue. Growing up, she burnt the skin off the palm of her hands on a pressure washer that her father had left running, and once ran around on a broken ankle for two whole days before her parents noticed the injury. She was once swarmed and bitten by hundreds of fire ants, has dipped her hands into boiling water, and injured herself in countless other ways, without ever feeling a thing.[10]

The experience of acute pain has been studied extensively using fMRI, generally by subjecting subjects to painful stimulation using a heat probe (or, in studies of visceral pain, a rectal balloon). This kind of acute pain results in activity in a broad set of brain areas that has come to be known as the "pain matrix"—which includes the somatosensory areas that receive sensory inputs from the body, as well as areas such as the insula and anterior cingulate cortex. You have already heard about these latter areas at many points throughout the book, so it hopefully occurs to you to ask whether activity in these areas is specific to pain, which in fact it is not. However, with the development of machine-learning techniques, researchers have begun to ask whether the experience of pain can be decoded from brain activity. Tor Wager is a researcher at the University of Colorado–Boulder who has led the charge to develop what he calls a "neurological signature of pain." In a landmark paper published in 2013, his group demonstrated that they could predict levels of reported pain with a high degree of accuracy using fMRI combined with a machine-learning technique that combined data in an optimal way across many different regions.[11] Given a person's brain image, they were able to predict that person's rating on a nine-point pain scale, with an error of about one point, and they were able to tell whether or not a person was in pain with an accuracy of greater than 90%.

What about different types of pain? Anyone who has experienced the heartbreak of a failed relationship will know that while the sensation is different from acute physical pain, it hurts nonetheless. In order to test whether their model could distinguish between physical pain and heartache, Wager and his colleagues recruited a set of subjects who had recently been broken up with. During scanning, they showed them pictures of their former lover as well as pictures of other friends, and also subjected them to physical pain in a separate scan. Although seeing pictures of the person who had rejected them caused activity in many of the same areas that are activated during the experience of physical pain, the algorithm was able to distinguish the two with a high degree of accuracy.

While acute pain is a useful signal to help prevent and limit injury, when it becomes chronic it can lead to misery, disability, and sometimes suicide. The community of cognitive neuroscience lost one of our brightest stars to chronic pain in 2012, when Jon Driver, a professor at University College London, committed suicide after suffering from chronic pain resulting from an injury sustained in a motorcyle accident the year before. Chronic pain is also a central feature in many civil law suits, in which individuals pursue damages related to their professed suffering. A challenge in these suits has always been that it was impossible to know whether the person is truly suffering from the claimed pain or is falsely claiming to be in pain in order to obtain a financial judgment or settlement. Could these new fMRI pain signatures provide a better way to validate claims of chronic pain? Potentially, but we are not there yet. Most importantly, chronic pain appears to have a different basis in the brain than acute pain. Research by Vania Apkarian and his colleagues has shown that chronic pain engages a different set of brain areas than acute pain, such that chronic pain involves areas that are more involved in emotional processing compared with the areas involved in acute pain.[12] This means that the neural pain signature developed by Wager's group would likely not work to detect many forms of chronic pain; a different tool would be necessary.

Just as in fMRI lie detection, the lack of a solid scientific background has not stopped people from trying to commercialize the techniques and use them in court. Unlike fMRI lie detection, pain detection *has* been allowed as evidence: in the case of *Koch v. Western Emulsions Inc.*, Carl Koch sued his employer over chronic pain resulting from an accident on the job, presenting fMRI evidence for the reality of his pain.[13] There are a number of companies that now market fMRI pain detection for the purpose of civil lawsuits in the United States. Most of these companies use secret methods, but one of them—Chronic Pain Diagnostics— has published its work in a peer-reviewed journal,[14] for which it should be applauded. We often think that publication in a peer-reviewed journal is a stamp of approval for the quality of a particular piece of research, but that's not always the case, as it depends on the peer reviewers having the right expertise to find flaws if they exist, which is especially tricky with new methods like machine-learning techniques. In the case of the work from the Chronic Pain Diagnostics team, the glaring flaw is that the sample size was far too small to make any meaningful inferences—only 13 people in each of the groups being studied. The study claimed to be able to detect chronic pain with 92% accuracy, but work by French machine-learning expert Gaël Varoquaux has shown small samples can lead to highly inflated estimates of accuracy on just this kind of test.[15] In addition to these concerns, questions remain about the potential ability for subjects to intentionally trick the scanner (which will also appear in the discussion of fMRI lie detection in chapter 6)—though research from Tor Wager's group has shown that the neural pain signature is not affected by imagined pain.[16] I certainly hope that one day fMRI will be able to help people in pain obtain the justice they deserve and prevent abuses of the legal system, but I think there is a substantial amount of hard work to do before we get there.

It is clear that fMRI decoding is already very powerful, and becoming more so as machine-learning techniques become increasingly powerful. In this chapter we have already seen a number of applications of fMRI decoding to real-world problems. In chapters 6, 7 and 8, we will dig further into the ways in which the use of decoding has started to impact business, medicine, and the law.

CHAPTER 5

HOW DO BRAINS CHANGE OVER TIME?

On September 24, 2012, I got into an MRI scanner at the University of Texas at Austin, where I was the director of the Imaging Research Center. It's not uncommon for MRI researchers like myself to get scanned; often when we are testing out a new technique we just need a warm body in the scanner to see if it works. But this was different, as it was the first of a large number of scans taking place over the course of more than a year. Over the next 18 months I would get into the MRI scanner 104 times at the University of Texas, as well as being scanned at Washington University in St. Louis and at Stanford. Why would anyone do this? For me it was a means to start answering a fundamental scientific mystery about how the brain changes over time.

When we think about how the brain changes, we first need to distinguish the time scale: Are we talking about changes over the course of years, weeks, or seconds? Second, we need to ask what is causing the change. Early brain development relies largely on our genome's plan for how to build a brain, but most changes in the brain rely upon an intimate interaction between our genes and our environment. In fact, every experience you have leaves an imprint (however tiny) on the structure and function of your brain, through a set of brain mechanisms known as "neural plasticity." The brain's mechanisms for plasticity are complex and still being heavily studied, but several facts are well established. When two neurons fire at the same time, the connections between them (that is, the synapses) get stronger,

such that the same amout of input from one neuron will cause
a stronger response in the other. This idea was first proposed
by the neuroscientist Donald Hebb in 1949, and for that reason
it is usually referred to as "Hebbian plasticity"—but a common
parapharasing is "neurons that fire together, wire together."
Neuroscientists now understand a great deal about the biology of
this kind of plasticity, including the molecules and genes that are
necessary for it to occur. We also know that this kind of plasticity
is necessary for learning; if we administer a drug to an animal
that blocks it, the animal's ability to learn will be diminished.
It is these plastic changes in neurons that underlie nearly all
of the ways in which we learn and remember, but *where* in the
brain the plasticity happens matters for what is being learned.
For example, our ability to remember events and replay them in
our mind (which psychologists call "episodic memory") requires
plasticity in a part of the brain called the *hippocampus*, while
learning a new motor skill involves plasticity in a different part
of the brain, called the *basal ganglia*.

Brain Development Across the Life Span

Our brains change continuously throughout our lives, from their
emergence during fetal development through old age when their
function often starts to decline. Interestingly, we are born with
nearly all of the neurons that we will have in our lifetime. It's true
that some parts of the brain can make new neurons throughout
our life and that this is very important for learning, but the
human newborn has nearly all of the 100 billion neurons that
it will have for the rest of its life. The connections between
those neurons, which are the key to learning, develop rapidly
up until about the end of the first year of life, when the brain
then starts to pick them off, through a process called *pruning*
that continues throughout one's life. However, neuroimaging has
shown us that this development doesn't happen equally across
the brain. Research by Jay Giedd, Elizabeth Sowell, and many
others has shown that the prefrontal cortex is the last part of the
brain to develop in terms of its structure, not reaching its mature
state until early adulthood, well after other parts of the brain.

Another aspect of brain development has to do with its wiring. I previously mentioned myelin in the context of imaging the white matter of the brain using diffusion weighted imaging. This fatty substance surrounds the neurons and helps them transmit information faster and more accurately. Research using diffusion weighted imaging has shown that the white matter develops very slowly, not fully maturing until well into the third decade of life. And just like the cortex, the white matter also doesn't develop evenly across the brain. The long tracts that connect the frontal lobe to the rest of the brain are some of the slowest to develop, while others reach their mature state by about 10 years of age. As I will discuss in chapter 6, the underdeveloped state of the prefrontal cortex and its wiring to the rest of the brain during adolescence provides part of the explanation for why teenagers can sometimes seem so out of control.

At the other end of the life span, a depressing reality of aging is that brain function starts to decline once we round our 30s. While some neurons die as we age, the bigger change comes in the connections between neurons that store our memories and knowledge, which start to deteriorate over time.[1] In addition, the myelin that insulates the brain's wiring also starts to degrade. These changes occur even in healthy aging, while another set of more insidious changes occurs in people suffering from age-related dementias such as Alzheimer's disease. In these diseases, the accumulation of damaging proteins in the brain's neurons causes them to start to malfunction and ultimately die. Some of the earliest changes happen in a part of the temporal lobe called the *entorhinal cortex*, which sends input to the hippocampus; this explains why memory problems are often the first sign of dementia, though it's important to note that a decline in memory function is a normal part of aging even for people without dementia. However, changes also occur throughout a network of regions in the cerebral cortex that are connected to the hippocampus. Research by Randy Buckner of Harvard and others combined structural MRI, fMRI, and PET imaging to show that there is a specific set of brain regions that is important for memory that starts to malfunction early in the development of Alzheimer's disease.[2] In particular, they used a newly developed type of PET imaging that allowed them to quantify the amount of the damaging protein known as amyloid throughout the brain. Using this technique, they were able to demonstrate that the network of regions involved in memory also had the highest levels of amyloid, as well as showing the greatest amount of atrophy. Buckner's work provides an outstanding example of how neuroimaging can provide new insights into brain diseases.

How Experience Changes the Brain

The brain can often repair itself in response to insults such as stroke or brain injury. It was once thought that the brain lost most of its plasticity after puberty, but we now know that the ability of the brain to change in response to damage extends well into adulthood. Remember the case of Lisa, who I introduced in chapter 1, who had undergone a radical surgery to remove her left hemisphere at age 16 in order to treat her severe epilepsy. She was a right-handed child, which means that her left hemisphere was almost certainly the dominant side for processing language; this is also consistent with the fact that she did not speak for about a year after her left hemisphere was removed. However, her brain retained a remarkable amount of plasticity, which was evident when we studied her several years later. While her language function was far from normal, she was able to read simple sentences aloud and have a basic conversation. To find the areas in her brain that were supporting this newfound language function, we scanned her with fMRI while she listened to words or made judgments about the meaning of printed words. In each case, what we saw was that there was activity in the areas of her right hemisphere that matched where we would have expected it in the left hemisphere of a healthy person. Somehow, over the course of those several years, her brain rewired itself so that her intact right hemisphere could take over the functions that had relied upon her left hemisphere before it was removed.

There is also evidence that experience across the lifetime can change the brain, even in the absence of brain damage or illness. One of the best test-beds for the study of these kinds of changes is in musicians, who usually spend many years learning and perfecting their complex combination of knowledge, auditory perception, and motor skills. Gottfried Schlaug from Harvard Medical School has spent more than two decades studying how musical experience changes the brain, and the results from his work as well as from others are very clear: musical experience changes both the structure and function of the brain, in direct relation to the amount of experience. One of the best-established findings is that the corpus callosum (the bundle of fibers connecting the brain's two hemispheres) is thicker in musicians compared with nonmusicians, and these changes are particularly evident in musicians who started learning music at an early age. This probably reflects the need for greater communication between the two hemispheres during musical performance, especially for players of keyboard instruments who have to coordinate their two hands. Other studies have looked at the size of the motor cortex, which in nonmusicians is usually larger on the dominant side (for right-handers that would be the left hemisphere, since everything is crossed over between the brain and the body). However, in pianists the motor cortex is much more equal in size across the two sides, primarily because it is larger on the nondominant side. It seems that the need to use both hands equally when playing the piano causes growth in the cortex that controls the nondominant hand, compared with nonmusicians who don't use the nondominant hand for fine motor skills as often.

The changes I have discussed so far happen over years or decades, but the brain's activity can also change much more quickly with experience. One of the most reliable findings in neuroimaging is that when a person does the same thing repeatedly, the amount of activity in the brain goes down.[3] A common example of this is a phenomenon known as "repetition suppression," in which repeatedly presenting a specific stimulus (such as a particular word or picture) leads to a decrease in the amount of activity in the areas of the brain that process the stimulus. For example, let's say that I present you with a noun (such as "hammer") and ask you to generate a verb that goes with the noun (you might say "hit"). The first time you do this, there will be activation in a set of networks in your brain that are involved in reading and language processing as well as executive control networks that are engaged more generally whenever you do a novel cognitive task. If I were to show you the same word again and ask you to do the same task, there would be much less activation across all of these networks (see color plate 7). This probably reflects a couple of different things. First, when you do the same task a second time, you will almost always do it more quickly, which means that these brain systems are turned on for a shorter period, leading to less overall activity when we average it over time. Research by Tal Yarkoni has shown that there is a large set of brain networks whose level of activity is correlated with reaction time regardless of the specific task that the person is doing, meaning that their activity probably indexes something very general about how difficult a task is.[4] Second, when we do something for a second time, we will often do it differently. The first time you generate a verb related to the word "hammer," you have to search your word knowledge to find the right word. However, if I ask you to do the same thing again a short time later, now you could either once again search your word knowledge, or you could simply remember that you said "hit" the last time you did the task. Because remembering the answer is often going to be faster and easier than searching through your knowledge base for an answer, it's generally going to evoke less activity in the brain.

Brain Fluctuations

The research that I just described has told us a lot about changes that happen very slowly (over years or decades) or very quickly (over minutes). My lab had worked on both of these topics for years, but at some point it started to occur to me that there was a serious missing link in our understanding of how the brain changes over time—namely, how does it change over the course of days, weeks, and months? This is important because it's exactly this time frame over which people with mental illness can exhibit huge fluctuations in their psychological function, which must reflect changes in their brain over that time scale. Detailed studies of individuals with schizophrenia and bipolar disorder have shown that their symptoms as well as their overall level of daily life functioning can fluctuate drastically from week to week: one week a person can be fully functional, and a couple of weeks later that person can be completely disabled.[5] Because we can't understand these kinds of changes in illness without first understanding how the healthy brain fluctuates over time, this seemed like a major void in our scientific knowledge—at that point there were no studies that had examined these kinds of fluctuations.

The reasons for such a scientific blind spot are not hard to see. First, cognitive neuroscientists have largely viewed the brain as a relatively static entity. We realize that the brain changes with experience, but nonetheless we generally assume that when we scan a person we are taking a representative snapshot of their brain function, and researchers have generally not thought that fluctuations over weeks or months are very interesting. Second, doing this kind of longitudinal research is really hard. Getting a person to come in for a single MRI scan is usually easy, but getting them to return for multiple scans over many months would be very difficult—just imagine that I asked you to come and get in an MRI scanner once a week every week for a year. In theory we could pay people enough to make sure that they come back for every visit, but in general the ethics boards that approve our research will not let us pay people an amount that they feel would be coercive; our subjects are supposed to be able to freely walk away from the research if they want to, but if they depend on the money to pay their rent or buy food then they don't really have that option.

Even if I could find willing volunteers, there is also the challenge of how to obtain research funding for such a study. In a perfect world, I would be able to apply for grant funding to study brain function in a group of volunteers over time, and given my track record I should have a good shot at receiving funding for such a study. However, most grant programs (such as those from the National Institutes of Health [NIH] and National Science Foundation, which fund most neuroscience research in the United States) will not support exploratory research where the answer is not obvious. Instead, this kind of research is often labeled pejoratively as a "fishing expedition"; in fact, even when I later applied for a grant to do this kind of work from one of the NIH programs designed to support exploratory new research, it was rejected because the reviewers thought that it was too exploratory. Thus I was stuck on the horns of a dilemma: I could not hope to obtain research funding to answer the question without having some preliminary results to provide hypotheses to test, but at the same time I could not secure the funding necessary to obtain those pilot data.

"My Crazy Study"

Just as I was starting to think deeply about the variability of human brain function over time, an unlikely source of inspiration drove me to begin thinking about using myself as the first test subject. Laurie Frick is a former high-tech executive turned artist, whom I met in 2011. After her interest in our research became clear, we appointed her as our imaging center's "artist in residence." This meant two things for us: First, she joined us for scientific discussions, often bringing an interesting outside perspective on the questions that we were asking. Second, she allowed us to borrow some of her art pieces, which helped make our austere scientific laboratory a lot more beautiful. Laurie was deeply entrenched in the Quantified Self movement, which is a group of people who record as much data as possible about themselves, and her art is based on data that reflect the patterns in her life or others' lives. As Laurie began to push me to take advantage of my unique position as an fMRI researcher to collect brain imaging data on myself, a paper appeared that would prove to be the added inspiration that I needed to begin seriously studying myself.

Michael Snyder, a molecular biologist from Stanford, published a groundbreaking paper in 2012, in which he described what has come to be called the "Snyderome" (or as the journal *Nature* jokingly referred to it, the "narcissome").[6] Snyder's lab at Stanford studies many of the different "-omes" that are central to modern biology: the genome that describes our genetic code, the transcriptome that describes how those genes are expressed, the proteome that describes the proteins generated from that expression, the metabolome that describes many of the small molecules involved in bodily metabolism, and more. His team developed an approach that they called "integrated personal-omics profiling," which involves quantifying almost everything that can be quantified about a human's biological function, and they used this to follow Snyder's personal biology over the course of more than a year. Analyses of Snyder's genome showed that he had genes that put him at risk for type 2 diabetes, and during the course of the study this genetic risk became destiny; after a respiratory infection, his blood glucose levels spiked and he developed full-blown diabetes. Now, no one wishes to develop a major disease such as type 2 diabetes, but Snyder's illness became a medical gold mine, because the blood collected over the course of his illness allowed his lab to generate the most detailed biological picture ever of how the disease develops. In particular, his results provided a set of new hypotheses about the relationship between inflammation and diabetes, which are now being tested in larger groups of prediabetic individuals.

Snyder is one of the world's leading molecular biologists, and his study showed me that serious researchers can make major discoveries by studying themselves. The idea of self-experimentation is of course not new; many researchers throughout history have experimented on themselves before studying others. In his memoir, Marcus Raichle recounts his experience as a young researcher when he spent several hours hyperventilating with catheters stuck in his jugular vein and femoral artery, in order to test how carbon dioxide affected brain metabolism. Self-experimentation has not always ended well (eight deaths of self-experimenters were recorded in the first half of the twentieth century),[7] but in some cases self-experimentation has led to discoveries that have changed the course of science and medicine. Take the case of stomach ulcers. It was long thought that these ulcers occurred as a result of stress or diet, but Dr. Barry Marshall had a theory that they could instead be caused by a bacterium called *Helicobacter pylori* (*H. pylori* for short). In order to establish that an organism causes a disease, researchers must show that introducing the organism into a healthy individual results in the disease, and that eliminating the organism cures the disease. Dr. Marshall established this by drinking a liquid containing *H. pylori*; soon thereafter he developed an ulcer, which was cured by a course of antibiotics that killed the bacteria in his stomach. This self-experiment has revolutionized the treatment of stomach ulcers, and gained Marshall a Nobel Prize in Medicine.

In early 2012 I began discussing my planned study with many of my colleagues, usually describing it as the "crazy study I'm thinking about," and in September 2012 I began the study. I started out with the plan of scanning myself three times a week. It was important that these scans be performed under the most controlled conditions possible, so we performed them at very specific times of day and days of the week—every Monday at 5.00 p.m., and every Tuesday and Thursday at 7.30 a.m. We also planned the collection of blood every Tuesday, so that we could perform some of the same kinds of -omics analyses that Snyder performed in his study. Because food can drastically affect these results, this meant that I needed to fast and avoid caffeine every Tuesday morning until after my blood draw.

One of the not-so-important questions we had to answer early on was what to call my study. Some of my collaborators on the project had taken to calling it either the "Russome" or the "Poldrome" but I didn't like either of those, partly because they seemed too self-aggrandizing but also because they didn't really highlight what was most important about the study. I settled on calling it the "MyConnectome study," riffing on Sebastian Seung's well-known TED talk titled "I Am My Connectome," which I mentioned in chapter 3. This name highlighted the fact that we were particularly interested in understanding how brain connectivity, by which we mean the relationships in activity levels across different brain regions measured using fMRI, changes over time within one person.

Soon after we began the study, a potential snag appeared. While planning the study, we had considered the potential side effects of repeated MRI scanning. Fortunately MRI does not involve ionizing radiation (like an X-ray or a CT [computed tomography] scan), and a large body of research suggested that the effects of repeated scanning should be minimal. The worst that I had expected was possibly some dizziness due to the crystals in my inner ear being pulled around by the magnetic field. However, in the first few weeks something worrisome happened. I suffer from tinnitus, or ringing in my ears, probably because of shooting guns without ear protection as a child (a hazard of growing up in small-town Texas) and too many loud rock concerts in my teens and twenties. It's like a little friend who is always there in the background, waiting to blow its whistle whenever I get bored or anxious. Soon after the study began, I noticed that my tinnitus became more intrusive. Anyone who has had an MRI scan will understand why—the scanner makes very loud noises, and the research scans used for my tests are even louder and more annoying. As much as I like scientific discovery, I like my hearing a lot more, so I quickly arranged to get tested at the campus hearing center, so that we could follow my hearing to make sure that I wasn't damaging it. The results of the first tests were not surprising—I had fairly significant hearing loss in the high frequencies, consistent with noise-related hearing damage, which predated my scanning endeavor. I continued to get tested monthly, and for many months the results were very consistent, but in April 2013 we noticed a worrisome reduction in my hearing. In retrospect this was probably a fluke (I had a bit of a cold on the morning of the test), but it concerned me enough that I decided to take a break from the study for a couple of months. Fortunately, follow-up tests later that summer showed that my hearing was unchanged from the beginning of the study, and we started back up with the study in June 2013.

I began the study with a plan of collecting one year's worth of data, but real life intervened in a number of ways, including work-related travel, holidays, MRI scanner problems, and a number of campus closures due to icy weather in the spring of 2014. I also had to stop doing the Monday afternoon scans, as they just took too much time out of my day. In the end, it took about 18 months to collect 48 samples of blood data (that's about a quart of blood in total) and 104 MRI scanning sessions. I moved to Stanford soon after the study ended, and I spent the summer of 2014 digging deeply into the data from a temporary apartment in Palo Alto, using the supercomputers at the University of Texas to process the massive data sets.

How One Brain Changes over Time

We performed many different types of MRI scans in the course of the project, but the one type of scan we focused on in particular was resting fMRI, which you have already encountered in chapter 3. The goal of resting fMRI is generally to understand the relationships between the activity of different regions across the brain over time. We record data from about 100,000 small cubes ("voxels") within the brain, but then we collapse the data from nearby voxels into regions (which we generally call "parcels"), through an operation that we call "parcellation." The idea behind parcellation is that there are regions of the brain whose connectivity with the rest of the brain is highly similar, so we can treat them as a single unit for the purposes of our data analysis. We worked with a group led by Steve Petersen at Washington University in St. Louis, which has developed some of the most state-of-the-art methods for parcellation of the brain. In fact, Petersen's group first contacted me because they had heard about the study and wanted to use my data set to test the reliabilty of their parcellation methods—they had never had enough data from one person to do so. This turned into a remarkably effective and enjoyable collaboration between our groups, spearheaded by a brilliant graduate student in Petersen's lab named Tim Laumann. Tim applied their methods to my data and found that they were indeed quite reliable; when he applied the same method to two different sets of scans collected on different days, the results came out very similar, giving us added confidence in the method. The parcellation results told us that my cerebral cortex was composed of 620 different regions. This was many more regions than previous research had suggested, but that work had been based on much less data for each individual. Subsequent research by David Van Essen and his colleagues in the Human Connectome Project has identified a similar but somewhat smaller number of regions (360) across many more people, using a different set of methods.

Once we had identified the 620 regions in my brain, we took the average activity from each region and performed all of our analyses on those data, which greatly reduced the computational burden of the analyses; instead of needing a supercomputer, I could now do most of the analyses on my laptop. We calculated the correlations between each pair of the 620 regions within each 10-minute-long resting fMRI scan, which gave us almost 200,000 correlations, which we then used to sort the 620 regions into a smaller number of "resting state networks"—-basically, sets of regions whose activity is more highly correlated with one another than with the rest of the brain. This showed that there were 13 networks, which corresponded quite well to the networks that the Petersen group had previously identified in larger groups of people. However, there were also some idiosyncratic features of my brain. For example, there is a network that is always found in the middle of the brain called the "default mode" network (shown in red in color plate 8), which was first identified by Marcus Raichle and his colleagues and which seems to be most active when a person is thinking introspectively, as we do when we are resting in an MRI scanner. My default mode area sits in the same place where we would have expected it to sit based on the previous studies. There is another network, called the "salience" network, that usually spans different parts of the frontal lobe and is involved in orienting to surprising things in the environment. I also have one of these (shown in light blue in color plate 8), and again it's mostly in the right place. However, if you look in the middle of the default mode area in my prefrontal cortex, you will see several blue areas surrounded by the red default mode areas—these regions are basically in the wrong place, at least according to what we expected based on studies of groups of people (with much less data for each person).

One of the other important questions that we wanted to answer with the study was this: How do the differences in resting state connectivity between days for a single person compare to the differences across people? If the differences in my brain from one day to the next were similar to the differences among people, then this would mean that we might not actually need to study individuals longitudinally over time—we could just compare different people. It turned out that the fluctuations in my brain from day to day were very different from the way that people's brains vary from one to another: in fact, when we looked at the regions that varied the most in their connectivity from day to day within my scans, those turned out to be some of the *least* variable among people. This is important because it tells us that if we want to understand the fluctuations in brain function over time within a single person, then we need to do the kind of intensive studies over time that I did in my study; we can't simply look across different people and ask how they differ from one another.

Because I needed to be fasted and caffeine-free on Tuesday mornings in order to get valid results from the blood analyses, we had a built-in experimental comparison of the effect of caffeine and food between Tuesdays and Thursdays. We have long known that caffeine affects blood flow to the brain, and some labs keep a coffee machine near the MRI scanner because of the lore that it produces better data when subjects are caffeinated during a scan, so I expected brain connectivity to be generally lower and more variable on days when I was uncaffeinated. When we analyzed the data we were surprised by the results. First, we found that overall connectivity was actually *greater* on days when I was caffeine-free. Second, when we looked at where this happened in the brain, we found that it was specific to a small subset of areas, which were some of the most primitive parts of the cerebral cortex: the visual areas that process basic visual input, and the sensory/motor cortex that is responsible for touch and movement. These areas were relatively unconnected when I was caffeinated, but became highly interconnected when I was fasted and uncaffeinated. It was almost as if, in its tired and uncaffeinated state, my brain moved into a more basic mode that focused on sensory input rather than higher-order cognitive functions. We assume that this is due to caffeine, but we can't rule out that it might also reflect the effects of food, since the two were always changed together.

Tim Laumann drafted the first paper describing our findings and we submitted it for publication in early 2015. The reviewers of the paper were quite enthusiastic, but also raised some questions about the analyses that we had presented. In the paper we had compared my brain with the data from a large group of individuals who had been scanned in less detail at Washington University. There were two major differences between how we had collected the data in Austin and how the data had been collected from the subjects at Washington University. First, they were scanned on a different model of MRI scanner than I had been. We don't generally expect different scanners to give us radically different results, but a large study called the Biomedical Informatics Research Network has shown that results can differ from scanner to scanner, so this was a legitimate concern. Another difference between the studies was subtle put potentially very important: in my scans I had kept my eyes closed, while the subjects in the Washington University studies had their eyes open, staring at a point on a gray screen. In order to address these questions, I flew to St. Louis on April 3, 2015, and spent about six hours lying in the MRI scanner at Washington University, doing resting fMRI scans with my eyes either open or closed, in order to address the questions raised by the reviewer (see figure 5.1). It turned out that having my eyes open did indeed result in a substantial difference in the connectivity of my brain, not just in the parts of my brain that process visual information, but also in the sensory/motor areas that showed such big effects of caffeine and food. This meant that any comparisons between the data sets had to be interpreted with that factor in mind. With this question answered, the paper was accepted and published later that year.[8]

What did we learn from all of this work? First, we now have an initial glimpse into how the function of a single human brain changes over time; it is not at all surprising that the brain changes over the course of weeks and months, but until our study it had not ever been measured. In addition, we have discovered some of the factors that cause brain function to fluctuate, including caffeine and food as well as my mood. We have also discovered that these differences in brain function within a person over time are not the same as the differences across people. This is essential, because it says that we need to study individuals over time more intensively if we want to

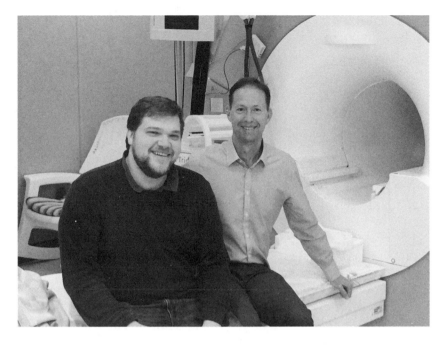

Figure 5.1. A photo of Tim Laumann (*left*) and me (*right*), just after I emerged from about six hours of MRI scanning at Washington University. Photo courtesy of Tim Laumann.

understand these fluctuations, which we think are crucial if we want to understand how the brain changes in mental illnesses such as depression or bipolar disorder. The results also showed us that if we want to obtain highly reliable measurements of an individual brain, we need to collect much more data about each person than most studies have done: behind the gross concepts of brain organization that had come from earlier studies (in which a small amount of data was collected from a large set of people) was hiding a level of fine-grained differences between people that have started to emerge from more detailed studies of individuals.

These findings have also inspired a number of other researchers to start looking at individuals in much more detail, and the findings have already started to change how we think about the functional organization of the brain. In one example of such a study, Rodrigo Braga and Randy Buckner from Harvard University scanned four individuals extensively and then examined how the default mode network in their brains was organized.[9] What they found was that within the areas where the default mode network should be found, based on the group studies, there were actually two separate networks next to each other, which differed in how they were connected to the rest of the brain. Because this organization differs from person to person, it had been lost in previous research that had combined data across individuals. The power of this "dense-scanning" approach is so strong that Buckner, one of the pioneers of fMRI research, has switched his entire research program to focus on the intensive scanning of individuals.

Toward a Personalized Neuroscience?

I am often asked what I learned about myself from all of the data that we collected as part of the MyConnectome study, and my answer is usually: "depressingly little." The data led us to some very important *scientific* findings, and those have been deeply edifying for me. It has also changed how I think about doing neuroimaging research, leading me to an interest in collecting more data on each individual in order to be able to better characterize individual brains. However, in terms of insights into how to live my life, there really have not been any. In part this is because while the project involved collection of an astounding amount of data, there is still too little data for me to confidently make decisions about how to change my life.

The most promising analysis for telling me something useful was what we called a "phenome-wide" analysis. The *phenome* refers to all of the different ways in which humans can vary, which come about through the interaction of our genomes and our environments. A phenome-wide analysis is one that examines the relationships between many different types of variables— in my case this included brain imaging, gene expression and metabolite levels in my blood, the foods that I had eaten each day, and all of the various psychological measurements that we made during the study. We examined the correlations between all of these different variables, which involved more than 38,000 statistical tests. After performing statistical corrections for this large number of tests, there were still a large number of statistically significant relations between variables. Some of these made very good sense; for example, the severity of my psoriasis (which I rated every evening) was related to the expression of genes related to T cells, which are known to play a central role in the disease. Others were potentially useful, such as a relationship between eating beef and the level of expression of genes related to inflammation. However, even though that relationship was statistically significant after correcting for the large number of analyses, it still only accounted for a few percent of the expression in those genes, and it's hard to know whether it's a false positive or a real result.

One central problem is that because this is correlational research, we don't know what causes what. Anyone who has ever taken a statistics class is familiar with the dictum that correlation does not imply causation. In reality, the presence of a correlation tells you that something is causing something else, but doesn't tell you which way the causal arrows point; in the case of the beef result, there are three possibilities. First, it could be that eating beef causes those inflammatory genes to be expressed more. Second, it could be that high levels of expression in those genes caused me to be more likely to eat beef (for example, as "comfort food" when I wasn't feeling as good). Third, something else could be causing both of these things to change. For example, perhaps I eat more beef when I am exercising more, and the latter also causes changes in gene expression. Because our data are purely observational, we can't tell these apart. For this reason, I didn't feel confident enough in any of the results to make big changes in my life (especially something as drastic as cutting beef from my diet).

In recent years there has been great excitement about the idea of "precision medicine," or "personalized medicine," with the NIH in the United States putting more than $50 million into a project called "All of Us," which is meant to gather data from more than one million people in order to more precisely treat their diseases. There are already some success stories for personalized medicine, such as the ability to effectively cure some cancers with treatments that are based on the specific genetic mutations in the cancer or the ability to tailor drug dosages based on a person's genetics. Much of the current excitement around precision medicine comes from the hope that data collected from mobile devices will provide physicians with new ways to understand disease and possibly intervene earlier in the course of a disease. For example, what if your smartphone could monitor your movements and use them to predict that you are about to become depressed? I also hold out hope that this kind of measurement will lead to more effective prevention and treatment of diseases. But my experience with the MyConnectome project suggests that it will be challenging to scale this up to studies that require more intrusive measurements, such as MRI scans. Even simple things like writing down all of the foods that I had eaten in a day became very tiresome after a few months, so anything that requires effort from the individual will likely be difficult to sustain unless it can be automated. I certainly hope that the All of Us project succeeds in showing new ways to prevent or treat diseases, but I think it's always important to disentangle hype from realistic expectations.

CHAPTER 6

CRIME AND LIES

Neuroimaging Meets the Law

When Christopher Simmons was 17 years old, he and a friend broke into a trailer home in Fenton, Missouri, in order to steal some money. They woke the trailer's resident, Shirley Ann Crook, whom Simmons recognized from an earlier car accident they had been involved in, and it was at this point that Simmons decided to murder her. He and his accomplice bound and gagged her with duct tape, drove her to a bridge over the Meramec River, and dumped her in the water, drowning her. Before Simmons was arrested, he bragged to his friends at school about the murder, and after he was arrested he quickly confessed to the crime and reenacted it on videotape. When his case went to trial, he was convicted of murder and sentenced to death by lethal injection. Twelve years later, the United States Supreme Court would rule in the case of *Roper v. Simmons* that giving the death penalty to someone under 18 at the time of the crime violates the Eighth Amendment of the United States Constitution, which prohibits "cruel and unusual punishment."

Neuroimaging research did not directly play a role in the *Roper v. Simmons* decision, but the case represented the first time that the Supreme Court used evidence from psychological research to determine that adolescents have an impairment (on average) in controlling their impulses, and thus cannot be held responsible for their actions. In subsequent cases, the court would go further and note that "developments in psychology and brain science continue to show fundamental differences between juvenile and adult minds. For example, parts of the brain involved in behavior control continue to mature through late adolescence," when they struck down life sentences without parole for juveniles convicted of nonhomicide offenses in the 2010 case of *Graham v. Florida*.[1] The nation's highest court has placed enough faith in neuroimaging research to base life-or-death decisions on it.

One of the basic challenges in bringing science to bear on legal questions is that science and the law have goals that are in many ways diametrically opposite. Both seek the truth, of course, but they seek different types of truths. Scientists usually aim to discover general laws that apply to entire populations: Does smoking cause cancer? Do adolescents have less ability to control their impulses than adults? Implicit in each of these questions is a coda that is usually left unsaid: "in general." The law, on the other hand, must make definite and final decisions about individual cases: Did smoking cause *this individual's* cancer? Did Christopher Simmons have reduced ability to control his impulses? Those are not questions that science can answer directly—at least not today, and probably never—though it can provide clues to how we might think about them.

Brain Development and Criminal Responsibility

One doesn't need neuroscience to realize that the adolescent brain is badly flawed. As William Shakespeare said in *The Winter's Tale*, "I would there were no age between sixteen and three-and-twenty, or that youth would sleep out the rest; for there is nothing in the between but getting wenches with child, wronging the ancientry, stealing, fighting."[2] But neuroimaging research has provided us with an increasingly clear picture of why adolescents act the way that they do.

We discussed in chapter 5 how the prefrontal cortex is slow to develop in children, and that it is not fully wired up to the rest of the brain until one's 20s. There is one brain system, on the other hand, that is particularly precocious in adolescents, which is the system that drives us to seek rewards. A central player in reward is the neurotransmitter dopamine, which we have already discussed in chapter 1. It's commonly thought that dopamine is a "pleasure molecule," but that's not quite right. The neuroscientist Kent Berridge of the University of Michigan has argued persuasively that we must distinguish between two aspects of reward: *wanting* (the desire to obtain something) and *liking* (the experience of enjoying something). Usually we experience these things together in our daily lives, but neuroscience has discovered that they actually rely on different brain mechanisms: dopamine is responsible for wanting rather than liking, while other systems (including the opioid and endocannabinoid systems) are responsible for the pleasure that we experience. More specifically, dopamine seems to play a role in "incentive salience," or the degree to which we are attracted to and motivated to obtain rewards in the world. Studies in rats also show that adolescent animals have higher dopamine release for rewards than younger or older rats. Unfortunately we don't have great data showing us how dopamine changes across adolescence in humans, because the only way to measure dopamine levels in a human uses PET imaging, and it is not ethical to expose children to radioactive materials just for the sake of research. We can, however, use fMRI to measure something that is a proxy for dopamine. The nucleus accumbens is a part of the brain that receives a very strong input from the dopamine neurons in the midbrain, and fMRI research has provided good evidence that activation in this area often reflects dopamine signals. The accumbens shows increased activity when a person receives an unexpected reward (which we know causes firing of dopamine neurons), and drugs that block dopamine also block these changes in the fMRI signal, which gives us fairly good confidence that they reflect dopamine, at least in part.

Research over the past decade has shown us that the nucleus accumbens is especially active during adolescence. Adriana Galván, now a faculty member at UCLA, discovered this in a study that examined how children, adolescents, and adults responded to pictures that were associated with different levels of monetary reward.[3] The adolescents showed greater activity in the accumbens than either children or adults when they were anticipating a big reward, which likely meant that more dopamine was being released. In later research, Jessica Cohen (at the time a graduate student in my lab) did a study that more closely linked the adolescent bump to dopamine.[4] She asked children, adolescents, and adults to perform a task where different images predicted different amounts of monetary reward, but not perfectly. Because of this, subjects sometimes experienced unexpected rewards (when the picture that usually came with no money suddenly came with a large amount), which we know should result in dopamine neuron firing (remember our discussion of "reward prediction errors" from chapter 1). What she found was that an area nearby the nucleus accumbens (which also receives heavy dopamine input) had an "inverted-U" relationship between age and activity for reward prediction errors; its activity for these prediction errors was highest in the adolescents compared with the children and adults, similar to what Adriana Galván had seen in her study. This increase in activity in brain systems for reward motivation, along with the relatively delayed development of the prefrontal cortex, helps explain some of the out-of-control behavior that we see in adolescents.

The bigger question that arises from these facts about brain development is whether they provide a reason to excuse adolescents from some of the responsibility for their actions. That is a legal and ethical question, rather than a scientific question, but science can help prime and test our intuitions. Take the case of a man described by Jeffrey Burns and Russell Swerdlow in the *Archives of Neurology and Psychiatry*:

A 40-year-old, right-handed man in an otherwise normal state of health developed an increasing interest in pornography, including child pornography. ... Throughout the year 2000, he acquired an expanding collection of pornographic magazines and increasingly frequented Internet pornography sites. Much of this prurient material emphasized children and adolescents. ... The patient went to great lengths to conceal his activities because he felt that they were unacceptable. However, he continued to act on his sexual impulses, stating that "the pleasure principle overrode" his urge restraint. He began making subtle sexual advances toward his prepubescent stepdaughter, which he was able to conceal from his wife for several weeks. Only after the stepdaughter informed the wife of the patient's behavior did she discover with further investigation his emerging preoccupation with pornography, and child pornography in particular.[5]

The individual was convicted of child molestation, but while waiting for sentencing he developed a severe headache and was taken to the hospital, where MRI discovered a massive tumor in his frontal lobe. The tumor was removed, and his sexual urges decreased; a year later they returned, and a follow-up MRI showed that the tumor had returned as well.

It's clear that the cause of this individual's abhorrent behavior was the tumor, since the urges appeared to track with the presence of the tumor, and I think many people would agree that this should absolve him of at least some moral and legal responsibility for his actions. In this case the line between tumor and no tumor is clear, but in most other cases it is not. Christopher Simmons was 17 years old when he committed his crime, and if he had committed the same crime a year later he would have been put to death with no questions regarding his brain's maturity, but from the case description it seems that he was always a particularly immature and impulsive person— so his brain at 18 would probably still be less well developed than some others who are much younger. In its appeals to the neuroscience of brain development, the US Supreme Court has taken a necessarily messy scientific landscape and drawn a bright line at age 18 when it comes to some kinds of punishments. Science can help us understand the situation, but ultimately the decision about where to draw the line on punishment is a moral, legal, and ethical decision, not a scientific one.

Lie Detection in the Courts

Nearly every criminal trial requires the court to decide whether defendants are honest and witnesses are accurate. The ability to detect lying and verify the accuracy of memories would be a major boon for the legal system, but lie detection and its use in law have been controversial for almost a century. In 1923,

James Alphonso Frye was on trial for second-degree murder, and tried to bring evidence of his innocence using an early version of the polygraph, which claimed to detect deception by measuring changes in blood pressure. In their decision in the case of *Frye v. United States*, a US Court of Appeals decided that this evidence was not admissible in court, stating:

> We think the systolic blood pressure deception test has not yet gained such standing and scientific recognition among physiological and psychological authorities as would justify the courts in admitting expert testimony deduced from the discovery, development, and experiments thus far made.[6]

This case would remain a standard for the admissibility of scientific evidence, until it was eclipsed by another case in 1993, which centered on whether a child's birth defects had been caused by a particular medication. In its decision on *Daubert v. Merrell Dow Pharmaceuticals*, the US Supreme Court outlined a set of criteria that any scientific evidence must pass in order to be admitted. Justice Harry Blackmun wrote for the court:

> The trial judge ... must make a preliminary assessment of whether the testimony's underlying reasoning or methodology is scientifically valid and properly can be applied to the facts at issue. Many considerations will bear on the inquiry, including whether the theory or technique in question can be (and has been) tested, whether it has been subjected to peer review and publication, its known or potential error rate, and the existence and maintenance of standards controlling its operation, and whether it has attracted widespread acceptance within a relevant scientific community.[7]

Since the Daubert ruling, there has been a rich debate on whether polygraph evidence meets the standards outlined in the ruling, with some courts allowing it despite the fact that the scientific community has largely concluded that the polygraph is not an accurate way to detect deception. In 2003 the National Research Council convened a panel to thoroughly review the scientific evidence in favor of the polygraph, and the verdict was resoundingly negative.[8] The report noted that while the polygraph can detect lying at levels better than random guessing, it is far from the level of accuracy that we require for confident legal decisions.

Can fMRI Detect Lies?

Just as the National Research Council was preparing its report on lie detection, Daniel Langleben of the University of Pennsylvania was working on a way to detect lies using fMRI. Langleben was studying children with attention deficit hyperactivity disorder (ADHD), specifically investigating executive control—the ability to control one's behavior based on intentions or goals, which is notoriously impaired in ADHD—and this work gave him an intuition that has proven to be central in the proposed use of fMRI for lie detection. He had noted that individuals with ADHD often have difficulty holding in the truth, as if they have an impulse to let it out, which inspired the idea that lying requires first inhibiting our natural urge to tell the truth. A number of researchers at that time were starting to study the brain systems involved in the inhibition of responses, which are spread across the prefrontal cortex, and Langleben thought that it might be possible to see activity related to lying in these brain areas. He devised a version of a test used in lie detection known as the *Guilty Knowledge Test*. Rather than an open-ended questionnaire like those used in standard polygraph exams, the Guilty Knowledge Test specifically probes whether a person has knowledge about a particular thing, such as details about the scene of a crime or the specific items that were stolen. Langleben created a version of this task in which volunteers were presented with playing cards, and asked to answer whether they had a particular card. However, they were told by the experimenter that they should lie about one particular card (the five of clubs), saying that they did not have it when they actually did. When Langleben compared brain activity for these "lie" cards compared with the other cards where the individuals told the truth, he saw activity in two different regions of the prefrontal cortex, showing that indeed there was a difference in brain activity for lying.[9] In a later paper, Langleben worked with an engineer,

Christos Davatzikos, to test whether they could decode whether the individual was lying on a trial-by-trial basis. They found that they could distinguish truth from lies in this situation with almost 90% accuracy, a seemingly impressive feat.[10] Around the same time, another group of researchers from the Medical University of South Carolina, led by Frank Kozel, performed a study in which they asked subjects to commit a mock crime, stealing either a ring or a watch, and found that they could decode which item had been stolen with 90% accuracy.[11] It was not long before there was a race to commercialize these techniques. One company, No Lie MRI, licensed the techniques that had been patented by Langleben and his colleagues, and began offering commercial fMRI lie detection services in 2006. Another company, Cephos, began offering these services in 2008, based on the techniques that had been patented by Kozel and his colleagues. At that point, it was just a matter of time until one of them ended up being tested in court.

The case of Lorne Semrau became the legal test-bed for fMRI lie detection. Semrau was charged in 2008 with fraudulently billing the US government about three million dollars for medical services provided by his company. He claimed that he was innocent, and as part of his defense he hired Cephos to perform fMRI lie detection on him in order to provide scientific evidence for his innocence. Without notifying the prosecution, Cephos arranged to scan Semrau, presenting him with questions regarding the alleged crime as well as a number of control questions. A first scan was performed on December 30, 2009, and the results were not exactly clear-cut—in one of the scans he had appeared to be deceptive, but this was attributed to fatigue. A follow-up scan was performed two weeks later, and on this one Semrau was judged to be telling the truth about the same questions he had failed in the first exam. Steven Laken, the CEO of Cephos who had performed the scans, told the court that Semrau's brain indicated he was telling the truth in pleading not guilty to defrauding the government.

When the defense attorneys tried to call Dr. Laken as an expert witness, the prosecution objected and called for a hearing to determine whether fMRI lie detection met the standards for scientific evidence laid out in the *Daubert* ruling. In addition to testimony from Steven Laken, the court also heard testimony from two experts who opposed the inclusion of fMRI lie detection evidence: the statistician Peter Imrey, and the neuroimaging pioneer Marcus Raichle. The first question that the court tackled was whether the error rate of the technique was low enough to be acceptable. Peter Imrey pointed out that the existing studies that had measured the error rate of fMRI lie detection had been done with very small samples, which meant that they were not very reliable. In addition, those error rates were based on studies of young adults in a research setting, and there was no way to know whether they could be generalized to real-world settings with an older individual (Semrau was 63 years old when he was scanned). The court then addressed the question of whether fMRI lie detection had been generally accepted by the scientific community. This was not a difficult call, given that there were a number of published papers by fMRI researchers (including Nancy Kanwisher) stating that fMRI lie detection was not ready for real-world application, and Marcus Raichle agreed. The final nail in the coffin was the fact that Laken would not definitively state that Semrau was telling the truth on any specific question, merely that he was telling the truth in general. Based on the discussion over those two days, the presiding judge Tu Pham decided that fMRI lie detection did not meet the standards of scientific evidence required to be entered into the trial, and Semrau was convicted of fraud. He appealed his case to the US Court of Appeals, arguing that the fMRI evidence should not have been excluded, but the Appeals Court agreed with the lower court on the exclusion, and the conviction was upheld.[12]

While the Semrau decision had a definite chilling effect on the introduction of fMRI lie detection into the courts, others have since tried to introduce fMRI lie detection at trial, with little success. A notable advocate for fMRI lie detection has been the celebrity physician Mehmet Oz ("Dr. Oz"), who has championed the case of Gary Smith, a retired US Army soldier and veteran of the Afghanistan war who had been accused of murdering his roommate and fellow veteran Michael McQueen in 2006. Smith was convicted of the killing, but the conviction was overturned on appeal. In his second trial, Smith's attorneys attempted to introduce fMRI data to provide evidence that he was not lying when he professed his innocence. A pretrial hearing was held by the trial judge to determine the admissibility of the fMRI data, with expert witnesses on both sides of the case (including neuroimaging experts Liz Phelps from New York University and Anthony Wagner from Stanford as witnesses against admission of the data). The judge ruled against admission of the fMRI data into trial, and Smith was convicted, but that conviction was also overturned on appeal, and Smith ultimately accepted a plea bargain. In an episode of his television show dedicated to the case, Dr. Oz interviewed Robert Huizenga, a celebrity physician who testified in the O. J. Simpson murder trial and starred as "Dr. H" on the television show *The Biggest Loser* and now runs the Truthful Brain Corporation, which had performed the fMRI scans on Gary Smith. Huizenga is breathless in his advocacy of fMRI lie detection:

> This is the first unbiased, scientifically backed way to differentiate a lie from truth-telling. There's 102 articles about it, over 500 authors, multiple machines ... it's very robust scientifically. I know you're a scientist at heart—this is not some fly-by gimmick. ... When you look at fMRI, you're going inside the brain—this is a paradigm shift in how we analyze truth telling![13]

Despite their setback in the Smith case, Huizenga continues to try to get fMRI evidence into courts, so far without success. However, his interests have also shifted, as he told me on a phone call that I made to check on the status of the Truthful Brain Corporation when its website was not accessible: "Sorry about the website being down, I've been working on human age reversal so I'm a bit swamped."[14]

What about those 102 publications that supposedly support the premise of fMRI lie detection? A search of PubMed, which is the authoritative bibliographic source for biomedical publications, listed 63 papers matching a broad search for "fMRI" and "lie detection" or "guilty knowledge detection" in June 2017, well after the claims by Huizenga in 2016 that there were more than 100 articles. Of those 63 publications, only 32 were actual research reports examining fMRI lie detection; the remainder were papers that simply mentioned in passing or discussed fMRI lie detection but did not directly study it. Most of the research reports were supportive of the premise of fMRI lie detection; however, several point out a number of problems, which calls into question Dr. Huizenga's claim to me in that phone call about those papers that "every freaking one of them says it works."

Will fMRI lie detection ever be good enough for the real world? It's certainly not impossible, but in order for it to work, there are several thorny issues that will have to be dealt with. Probably the most important and most difficult is the need to quantify its accuracy in realistic situations. Currently the data on accuracy of fMRI lie detection come almost exclusively from studies of healthy young adults who are lying because they were told to do so by the experimenter. There are several aspects of this situation that are unrealistic in comparison with the court. First, the stakes are low for the research subjects (usually just a small amount of money, if anything at all), whereas they can be life changing for the criminal defendant. Second, the research subjects know that the experimenter is expecting them to lie, and thus may be less motivated to hide it compared with a criminal defendant. Third, the research subjects have usually just encountered the lie in the context of the laboratory for the first time, whereas by the time of a trial a criminal defendant may have been rehearsing the lie for many years. This concern is apparent in the advice that George Costanza gave to Jerry Seinfeld about how to fool a polygraph test: "Jerry, just remember. It's not a lie if you believe it."[15]

It's also possible to fool fMRI lie detection using countermeasures. A simple countermeasure would take advantage of the fact that holding one's breath or shaking one's head causes massive changes in fMRI signals, much bigger than any changes caused by a task. Thus, doing these things at different points throughout the scan would introduce so much noise as to cause the data to be worthless. However, these countermeasures would also be easily detected by the examiner. A number of other strategies would be less easily detected, but also effective. A study by Giorgio Ganis and his colleagues in 2011 showed that while they could detect deception about the subject's birth date with perfect accuracy, the accuracy dipped to 33% when subjects performed a simple countermeasure of imagining a particular finger movement when they saw a date other than their birthday.[16]

Finally, there are some uncomfortable questions that must be discussed regarding conflicts of interest. Several of the researchers who have been involved in studies of fMRI lie detection have patented their techniques and licensed them to companies, making unspecified amounts of money in exchange for use of this intellectual property. Whenever researchers stand to gain financially from a scientific idea, there are legitimate concerns about their ability to impartially test those ideas. All scientists tend to become advocates for their own theories and ideas, but the presence of financial incentives always raises concern that researchers will look past negative evidence lest it damage their pocketbooks. In the case of lie detection, there have been a few studies by individuals without conflicts, but these have been rare. In the future, it will be important for studies of fMRI lie detection to be performed by impartial researchers using methods agreed upon by both the proponents and the critics. Only in this way will we be able to achieve the kind of evidence that the courts need to make an informed decision about the relevance of fMRI lie detection.

Predicting Future Crimes

Philip K. Dick's novel *The Minority Report*, made into a blockbuster film by Steven Spielberg, envisioned a world in which crime could be predicted and prevented through psychic powers. It seems unlikely that we will ever be able to predict crime with such pinpoint accuracy in reality; even if we could read criminal intentions from someone's mind, many crimes occur in the heat of the moment, without any premeditation. However, it would not be surprising if we could predict how likely an individual is to commit a crime in the future: if a person has committed a crime in the past, that person is much more likely to commit one in the future, and men are more likely to commit crimes

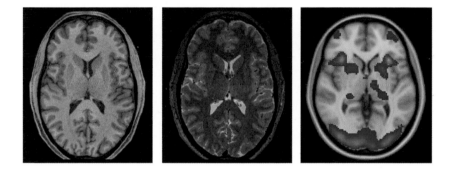

Plate 1. Magnetic resonance imaging (MRI) can be used to measure many different aspects of brain tissue. The *left two panels* show two examples of different types of structural MRI, which measure different aspects of the tissue such as how much water or fat is present. The *right panel* shows an example of a functional MRI activation map—the areas shown in red are those that have increased activity during a particular mental task. These "activations" are overlaid on a structural MRI image to show their locations in the brain.

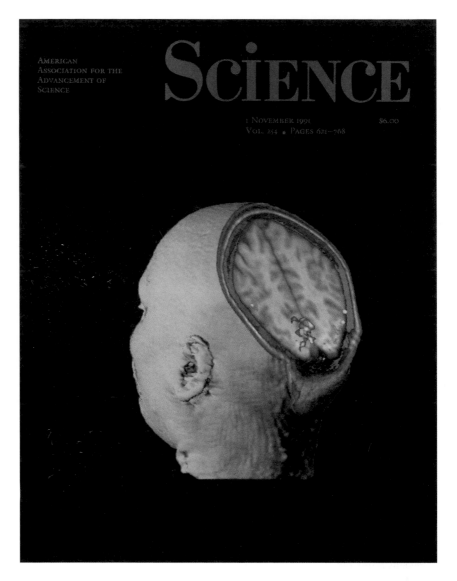

Plate 2. The cover image of *Science* from November 1, 1991, showing activation of the visual cortex (shown in red/yellow) as measured by Belliveau and colleagues. Belliveau et al. used an earlier version of functional magnetic resonance imaging (fMRI) that utilized an injected contrast agent. From *Science*, November 1, 1991. Copyright © 1991 by AAAS. Reprinted with permission from AAAS.

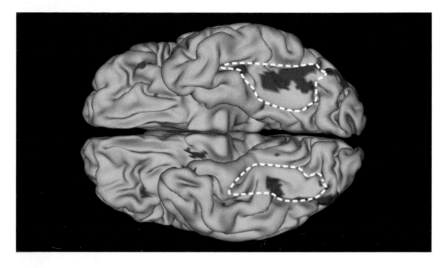

Plate 3. A view of my brain from the bottom, showing in red areas of my brain that were active when I viewed faces. The activity follows along the fusiform gyrus on each side (which is outlined with a dashed white line).

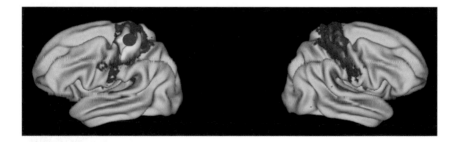

Plate 4. One way to analyze connectivity in the brain is to look at how the activity of the entire brain is related to a particular location, or seed. This figure shows a map of regions across the brain whose activity (shown in yellow and red) is correlated with that of a seed region (shown in blue) that is located in the motor cortex. This analysis reproduces the original findings by Biswal, which showed that the motor cortex on the opposite side of the brain was correlated at rest. Generated using data obtained from the Human Connectome Project.

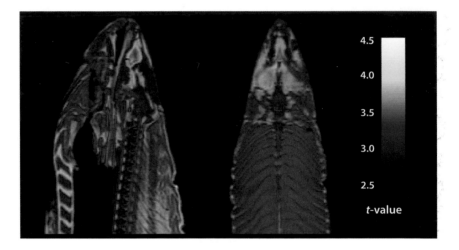

Plate 5. The image of brain activation in a dead salmon from Craig Bennett's 2009 poster. The tiny red spot was the location of significant activation found when the proper correction for multiple comparisons was not applied. Image courtesy of Craig Bennett.

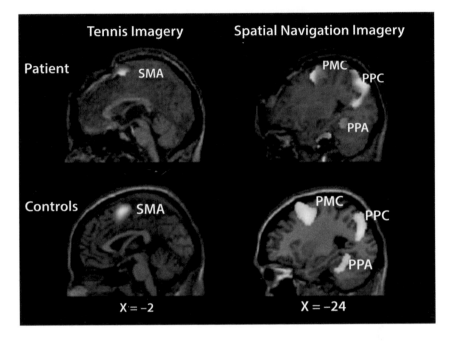

Plate 6. The results from Adrian Owen's 2006 paper, showing in red/yellow the brain areas active while imagining playing tennis (*left*) and imagining navigating one's house (*right*) in the vegetative patient (*top*) and healthy controls (*bottom*). From "Detecting Awareness in the Vegetative State" by Adrian M. Owen, Martin R. Coleman, Melanie Boly, Matthew H. Davis, Steven Laurys, & John D. Pickard, Science, 08 September 2006: 1402. Copyright © 2006 by AAAS. Reprinted with permission from AAAS.

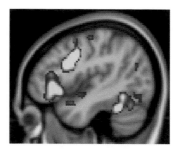

Plate 7. A map showing brain areas in the left hemisphere whose activity was greater the first time subjects decided whether a word was abstract or concrete, compared with the second time. The areas shown in red/yellow showed a statistically significant reduction in their activity when a word was repeated. Generated using data from the OpenfMRI project.

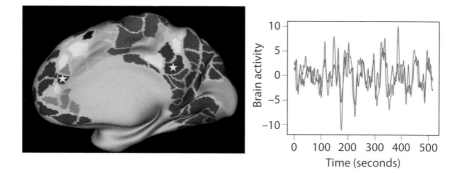

Plate 8. Mapping networks in the brain. The *left panel* shows the regions, or parcels, identified in the right hemisphere of my brain, looking at the middle of the hemisphere. The plots in the *right panel* show the fMRI signal in two of the parcels that are part of the default mode network (marked with stars). Despite being at different ends of the brain, these regions show activity that fluctuates in a very similar manner across the 10-minute resting fMRI scan.

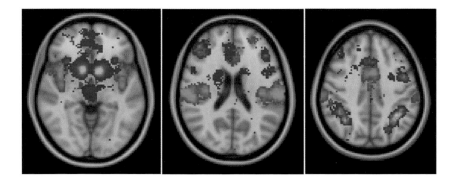

Plate 9. Results from three meta-analyses performed using Neurosynth. The three sets of results are shown together; areas shown in red are those where activation was associated with working memory across studies, green was associated with pain, and blue with reward. These results show that different psychological states engage reliably different patterns of activity across the brain.

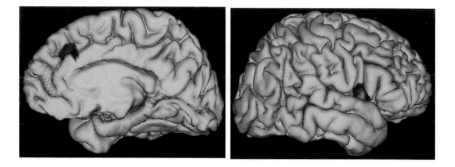

Plate 10. The areas (shown in red) found by Etkin and colleagues to show differences in brain structure related to mental illness that were common between different diagnoses. The *left panel* shows a view of the middle of the brain, highlighting the anterior cingulate, and the *right panel* shows the side view, highlighting the insula (though most of the insula is hidden away within the folds of the brain). Generated using data generously provided by Amit Etkin.

than women. In fact, such predictions are already part of many criminal proceedings, based either on the clinical judgment of an interviewer or on the use of data about the individual, such as age and criminal history (which is called "actuarial" prediction). These factors are used to decide whether individuals should be allowed out of prison, and whether they require additional supervision after their release. A large body of research has shown that actuarial prediction consistently outperforms the judgment of expert clinicians for the prediction of future violent behavior. For example, a recent study of individuals released from Swedish prisons showed that it was possible to predict who would be rearrested in the following two years with a positive predictive value of 75%; that is, when the actuarial model predicted that a person would be rearrested, it was correct 75% of the time.[17] Its negative predictive value (that is, how accurately the model could predict that someone would *not* be rearrested) was 65%. These levels of accuracy are good compared with guessing, but nowhere near the level of accuracy we would like for making decisions about whether someone should be free or remain in custody. Can neuroscience do better?

Kent Kiehl certainly thinks so. Kiehl has been called a "psychopath whisperer" because he has studied the brains of criminals and psychopaths for more than 20 years in the hopes of understanding what aspects of brain function cause people to commit antisocial acts. Because it would be difficult to bring incarcerated individuals to an MRI research center, Kiehl takes the scanner to the prison, using an MRI scanner housed in a tractor-trailer. In 2013 Kiehl and his colleagues published a study that seemed to herald a new era in the use of neuroscience for predicting criminal behavior. In a paper titled "Neuroprediction of Future Rearrest," they described a study in which 96 individuals were scanned in Kiehl's mobile MRI lab upon being released from a New Mexico prison, and then followed for four years to see whether they were rearrested.[18] In the scanner, the subjects performed a "go/no-go" task, in which they had simply to press a button when they saw the letter "X" (which occurred on the majority of trials), but do nothing when they saw the letter "K" (which occurred only rarely). This task requires inhibiting the usual response on those few trials where the K appears, and many psychologists (including Kiehl and myself) think that this kind of inhibition is probably one of the basic psychological processes that is deficient in people who commit crimes. About half of the inmates had been rearrested after four years. When Kiehl and his colleagues compared brain activity where people had failed to stop their response to a K to activity where they had correctly withheld their button press, they found that people who had been rearrested showed less activity in a specific place in the brain (the anterior cingulate cortex) than those who had not been rearrested. In fact, when they split the group in half according to its response in that area, they saw that the people with low activity in that region were rearrested 60% of the time, whereas those with high activity were rearrested only 46% of the time.

When I first saw this paper, a critical flaw jumped out at me. The authors used the term "neuroprediction" to describe their results, but in fact their estimates of prediction accuracy were compromised because they used the same data to create the statistical model and to test it. Statisticians have long understood that when one does this it leads to overly optimistic estimates of how well the model would work if it were then applied to a completely new data set. Because Kiehl and his colleagues had (very generously) shared their data along with the paper, my statistician colleague Jeanette Mumford and I were able to download the data and test how well one could actually predict on new individuals, using the cross-validation technique that I discussed in chapter 3.[19] For each person in the sample, we left one particular person out and fit a statistical model using all other individuals, and then tested the model's prediction for that left-out person. We did this using a baseline model that just included the person's age, as well as with a model that included both age and brain activation measure. What we found was that including the brain activity measure did indeed increase our accuracy at predicting whether someone would be rearrested, but only by about five percent. This is big enough to be of scientific interest, but it certainly is not big enough to warrant using fMRI for criminal prediction in the real world.

Does this mean that neuroimaging data are never going to be useful for predicting future criminal activity? Of course not. It is perfectly plausible that one day neuroimaging will be a useful addition to the existing models for risk prediction, but there are a few difficult challenges that will need to be addressed. First, just because brain activity is correlated with criminal activity (which is all that neuroimaging can show), that doesn't mean that it is the ultimate cause of criminal activity. What if childhood adversity and poverty lead both to a higher likelihood of criminal activity and to difference in brain activity (say, owing to differences in educational experiences)? The presence of a correlation doesn't tell us what is causing what. Second, we have to keep in mind that crime is a complex sociocultural phenomenon, and even if we can find a biological correlate that doesn't mean that the answer is a biological one. There is an ongoing debate around the use of actuarial risk prediction tools (like the Swedish one that I mentioned above), because they include variables that may be related to race or social class, such as disposable income.[20] If these methods of prediction are used to give preferential treatment to some offenders, then they could perpetuate or even exacerbate the societal inequalities that may have led to those differences in crime in the first place.

Is fMRI Reliable Enough for Real-World Decisions?

A final concern centers on whether the results from published fMRI studies are reliable enough to use them in ways that will deeply impact people's lives. Most people assume that when a scientist reports a research finding that it is likely to be true, but this assumption has come under intense scrutiny, owing in large part to the work of a researcher named John Ioannidis.

In 2005, Ioannidis was a researcher at a small university in Greece, having completed his medical training at Harvard and Tufts and established himself as a medical researcher studying the treatment of HIV infection. Over time, his interests had turned from clinical research to what he now calls "meta-research"—that is, the study of how scientific research is done. For the previous decade he had become increasingly concerned about problems with how medical research was done, which too often led ultimately to "medical reversals," in which the field of medicine suddenly decides that its practices have been all wrong.[21] During a trip to the small Greek island of Sikinos, he began to write the paper that would become his calling card, titled "Why Most Published Research Findings Are False."[22] In this paper, Ioannidis outlined a theoretical model for how scientific decisions are made and the factors that could lead them to be true or false. His analysis focused on a statistical concept of the "positive predictive value," which as I mentioned earlier in the chapter is the likelihood that a positive result found by a researcher is true. Let's say that I am a medical researcher studying whether a new drug improves symptoms in people with multiple sclerosis better than existing treatments. In the best case I would use a *randomized controlled trial*, in which patients are randomly assigned to receive either the drug being tested or the standard treatment for the disease, and their symptoms are measured and then compared statistically. Let's further suppose that our analysis finds a statistically significant difference between the new drug and the standard treatment. The positive predictive value reflects how likely it is that this is a true positive rather than a false positive—that is, whether the new drug is really better than the old drug, or whether we have made a statistical mistake. In his paper, Ioannidis outlined a set of factors that can decrease the positive predictive value of a finding, and argued that in general the positive predictive value of research findings is much lower than most of us would like to think.

Perhaps the most important of these factors is how much *statistical power* the study has. Power is a statistical concept that refers to how likely we are to find a true effect if it really exists, which depends both on how large the sample size is for the study and how small the effect is that we are trying to study. A very large effect can be found even with a relatively small sample—for example, smokers have an incidence of lung cancer that is more than 20 times that of nonsmokers, so we don't need to study huge populations to observe a difference in the rates of lung cancer between smokers and nonsmokers. However, many of the effects that are investigated in biomedical research are much smaller than this—for example, overweight individuals are 1.35 times more likely to have heart disease compared with normal weight individuals. That is still a fairly important effect, but it requires a much larger sample size to detect such an effect with confidence. Until Ioannidis's 2005 paper, most researchers had focused on how the power of a study is related to false negatives—everyone realized that smaller studies are less likely to find an effect even when it truly exists. What Ioannidis pointed out was that the power of a study also affects the positive predictive value. Think of it this way: if researchers perform a study with zero power that means that they have no chance of finding a true positive effect even if it exists. However, remember that there are always going to be a number of false positive results—how many being determined by the false positive rate that we specified in our analysis (usually five percent). So, in the case with zero power, we will have zero true positives and five percent false positives, meaning that the positive predictive value—the proportion of all positive results that are true—will be zero percent! As the power of the study increases, we start finding more true positives alongside our constant five percent of false positives, so the positive predictive value goes up.

Sometimes scientific papers can take a while to have their full effect, as if the world needs time to catch up with them. When Ioannidis's paper was published in the journal *PLOS Medicine* in 2005, there was some initial controversy, but the paper did not have a broad impact immediately. However, by 2011 it was becoming increasingly clear that Ioannidis's claim that "most published research findings are false" might actually be correct, at least within the field of social psychology. There were several intersecting causes for this concern. One was the discovery that Diederik Stapel, a prominent social psychologist, had published a number of high-profile papers based on data that he had faked. The fact that these claims had not been challenged by other researchers, despite the apparent rumors that his results were not replicable, led to concern about the reliability of other research in the field. Another source of concern arose around a paper published by the social psychologist Daryl Bem, which had claimed to find scientific evidence for precognition (that is, the ability to see the future). Statisticians who dug into Bem's results realized that they had probably arisen not because he had truly found evidence for the paranormal, but because he had tortured his data in a way that allowed him to find false positive results that fit his hypothesis. From this, the term "p-hacking" was born, referring to the fact that researchers can and sometimes will run many different analyses in an attempt to find a statistically significant result. Uri Simonsohn and his colleagues, in a paper with the provocative title "False-Positive Psychology," showed that when researchers take advantage of the various kinds of flexibility that are present in research (such as stopping the study as soon as one has found a statistically significant effect), almost anything could be found to be statistically significant.[23] In their example, they showed that using these methods they could find statistical evidence for the nonsensical conclusion that listening to a particular song (the Beatles' *When I'm Sixty-Four*) made people more than a year younger compared with a control song. The economists Leslie John, George Loewenstein, and and Drazen Prelec came up with a more general term for this kind of analytic trickery: "questionable research practices." They conducted a survey of a large number of psychologists (which I remember taking), and the results were striking: many researchers admitted to research practices that were very likely to increase the prevalence of false results. While almost everyone agreed that falsifying data was wrong and only about one percent of researchers admitted to having done it, there were other practices that were surprisingly common amongst researchers, such as deciding to collect more data based on whether the results were statistically significant (which more than half of researchers admitted to) and reporting an unexpected finding as if it had been predicted all along (which about a third admitted to).

The crisis came to a head around the inability of independent researchers to replicate a number of prominent findings in social psychology.[24] The social psychologist Brian Nosek put together a major effort to determine how many of the reported results in the psychological literature are actually reproducible.

He organized more than 250 other researchers to join him in attempting to replicate 100 published research studies, in an effort known as the Reproducibility Project: Psychology (RPP for short). The results, published in 2015, were shocking: whereas 97% of the original papers had reported statistically significant findings, only 35% of the replications found significant effects.[25] John Ioannidis's predictions had been borne out, though this did not give him pleasure. As he told the journalist Ed Yong: "The success rate is lower than I would have thought. ... I feel bad to see that some of my predictions have been validated. I wish they'd been proven wrong."[26]

If psychological research is as unreliable as the RPP suggests, then what about neuroimaging? Because neuroimaging research is so much more expensive than psychological research (often costing well over $25,000 to run a single fMRI study), an effort on par with the RPP would be almost impossible. There are a couple of reasons to think that many of the conclusions from neuroimaging might be relatively reliable. First, many of the very basic findings are visible in individual subjects assuming that enough data are available; nearly every healthy human has a face-responsive area in the fusiform gyrus and a default mode network. Second, studies that combine data across many different research studies (known as *meta-analyses*) show very consistent patterns of activity in relation to specific mental processes such as language or social function (see color plate 9), and many of these fit with other evidence such as the effects of brain lesions. Finally, using large data sets such as the Human Connectome Project we can analyze how well the results overlap across large groups of subjects, and we find that they are generally quite reliable.

At the same time, there are very good reasons to think that a substantial number of findings from neuroimaging research may be false. A major reason is that many neuroimaging studies have very low statistical power, which makes any positive results more likely to be false. This first came to light in a paper published by Kate Button and colleagues (including John Ioannidis), titled "Power Failure."[27] Button and her colleagues analyzed studies from across neuroscience, including both neuroimaging studies (though not fMRI studies) and studies in rats. What they found was that these studies overall had terrible statistical power; whereas we usually shoot for 80% power, meaning that we have an 80% chance of finding a true effect if it exists, many studies in neuroscience had power of less than 10%. With a number of collaborators I did a subsequent study of statistical power specifically for fMRI studies, and found that most of these studies were also badly underpowered.[28]

Where does this leave us? I think that we have to be very careful in our interpretation of published fMRI research studies. There are a number of questions we have to ask about any particular study. First, how large is the sample size? There are no hard and fast rules here, because the necessary sample size depends on the size of the effect that is expected; very powerful effects (like activity in the motor cortex when someone makes a movement) can be found with small samples, while weaker effects (such as those involving correlations between brain and behavior across people) require much larger samples, usually 100 or more; the tiny sample of 16 in our 2007 paper just doesn't make the grade. Researchers can use a technique called "power analysis" to find out how big a sample they need to find their effect of interest, and this should be the gold standard for determining the sample size for a study.

A second question is whether the analyses and hypotheses were planned before the study was performed, and whether these plans were followed. One of the developments arising from the reproducibility crisis in psychology is the idea that the methods of a study should be "preregistered"—that is, a description of the methods should be deposited in a database where they will be available to anyone once the research is complete, so that we can see that the methods were actually planned in advance as opposed to reflecting p-hacking or other questionable practices. At one point this would have required mailing a letter to one's self, but now there are websites that provide the ability to register and time stamp a hypothesis. This idea has been used for clinical trials in medicine for more than a decade now, and while there are still problems, it has helped improve the reliability of clinical trial research, mostly by reducing the number of positive outcomes, some of which were presumably false positives. However, to paraphrase Helmuth von Moltke, no analysis plan survives contact with the data—once we start analyzing the data we often realize that there were issues that we had not envisioned when we first planned the study. One important point is that deviating from the planned analysis is acceptable as long as one is transparent about the deviations and the reasons for them. Ultimately we would also like to have a separate and independent data set that we can use to confirm the finding; this is now the standard practice in some other fields of research, such as genetics.

Finally, we need to remember that science is a process for gaining understanding, not a body of knowledge. We learn from our mistakes and we move forward, with the realization that all of our knowledge is tentative and will likely be revised or overturned in the future. The willingness to change our mind and, indeed, our efforts to find our shortcomings and address them are the hallmarks of science that drew me to it originally and give me continued faith that it can help us better understand the world.

CHAPTER 7

DECISION NEUROSCIENCE

Imaging the Brain's "Buy Button"

Angelina Jolie announced in the *New York Times* in 2013 that she had chosen to have an elective double mastectomy, even though she did not have cancer. She made this choice because of her family history of breast cancer (her mother had developed the disease in her 40s and died of it at 56) and because she carries a mutation in a particular gene (*BRCA1*) that puts her at high risk of breast cancer. Her doctors had estimated that she had an 87% chance of developing the disease at some point in her life. On the other hand, the surgery carries a small risk of death (less than 1%), and it's also possible to have complications that could reduce one's quality of life. What would you do in this situation?

Every day we make thousands of choices. Some of them are inconsequential—should I eat yogurt or eggs for breakfast this morning? Others are not very important in the short term, but might have consequences in the long term—should I take the stairs or the elevator to my office on the third floor? Others are life altering, like Jolie's choice to undergo major surgery now in order to avoid an uncertain risk of disease in the future. For many years, the way that we make these choices has been studied by economists and psychologists, who have developed powerful theories that can describe many aspects of how we make decisions. In the past two decades, a new field called *neuroeconomics* has emerged that has attempted to describe how the brain makes these decisions, and neuroimaging has played a central role in the development of this new field.

Why Do We Choose What We Choose?

Let's say that I were to offer you a 50/50 chance to either win $25 or pay $13—would you take the chance? Most of the research into how we make decisions has studied these kinds of economic decisions, because they are relatively easy to study in the laboratory and because they are so important in our daily lives in modern society. The first aspect of an economic choice is the value of the different items being offered (which we call *prospects*). "Value" is one of those terms that we all intuitively know the meaning of, but providing a formal definition has long been a major challenge. We don't mean the numeric value or price; rather, we mean how much benefit you expect to get personally from the object or how much you want it, given what you already have. Economists came up with the term "utility" to describe the hidden quantity that drives us to want some things more than others. They tend to assume that while we can't measure this utility directly, we can infer it from the choices that a person makes; if I choose yogurt over eggs for breakfast, that implies that yogurt has greater utility for me than do eggs. If I am a rational decision maker, then I should always choose the outcome that has higher utility, because this should in theory satisfy me the most.

Another important aspect of a choice is how likely each of the outcomes is to occur. You would probably like a chance to win a new car, but you will be much more excited by a 50% chance than you will about a one-in-a-million chance. We can put together the probability and the face value of a prospect to compute the "expected value"—that is, how much would you expect to get on average if you took the gamble many times. For example, take the roulette wheel. A simple bet is that the ball will land on a specific square, which has a 1 in 38 chance of happening on an American roulette wheel, meaning that if you were to play many times you would expect to win about 2.6% of the time. If you win the gamble, you get paid 35 times the amount you laid down (plus keeping your original bet). If we multiply the probability of winning by the amount that can be won betting $100, it comes out to about $95; that is, if you were to play roulette many times, you should expect to come out around 5% worse off than you started. The fact that casino games are rigged against the player should not be surprising to anyone—so why do people flock to casinos in such large numbers?

The realization that people don't make choices based simply on expected value is not new, and in fact it was first discussed by the mathematician Daniel Bernoulli in the 1700s. Bernoulli was particularly interested in the fact that we often get less and less pleasure as we get more and more of something; you might be willing to pay $2 for a candy bar, but you likely wouldn't want to pay $20 for 10 of them. This idea that utility decreases the more we have can explain some of the ways in which humans seem to be irrational. For example, say that a particularly generous billionaire were to give you a choice between two options: a 50/50 chance to win $10 million, or $2 million for sure. Very few of us would pick the uncertain gamble over the sure $2 million, even though the expected value of the gamble ($5 million) is much greater than that of the sure thing. Bernoulli's explanation for this would be that as the amount of money goes up, the amount of added pleasure (or "utility") that we get from each dollar decreases, so that the difference in pleasure between nothing and $2 million is greater than the difference between $2 million and $10 million. In fact, sometimes utility can actually go down with more money—as explained by the hip-hop artist The Notorious B.I.G.: "It's like the more money we come across, the more problems we see."[1]

A deep understanding of the ways in which humans often behave irrationally did not come about until the work of Israeli psychologists Daniel Kahneman and Amos Tversky.[2] Kahneman and Tversky demonstrated many new ways in which human behavior is at odds with the economic concept of the rational decision maker and developed a theory (called *prospect theory*) that remains one of the most powerful theories in the study of decision making. Kahneman won the Nobel Prize in Economics for this work in 2002; Tversky would have shared it with him had he not died of melanoma in 1996 at the age of 59. Prospect theory takes the idea of expected utility but then adds several wrinkles, two of which are relevant to our discussion. First, they pointed out that the psychological impact of a loss is greater than the impact of the same amount of gain; losing $20 hurts more than gaining $20 feels good, a phenomenon they called "loss aversion." This explains why most of us would not accept a gamble that offered a 50/50 chance to either win $12 or lose $10; even though the expected value of the gamble is positive, the impact of the loss is magnified compared with the gain, by about a factor of two for the average person. Second, they demonstrated that people will often behave differently depending on how a choice is described (or, in their terminology, how it is "framed"); when an outcome is described in terms of avoiding a loss people are more likely to take risks, compared to the same outcome described in terms of a gain. This theory has had a major impact on economics and psychology, and, as we will see shortly, on neuroscience as well.

The Neuroscience of Choice

The ease with which we make choices in our daily lives belies the incredible complexity of the computations that our brain must perform to make those choices. However, a quick look at the problems in decision making that occur with brain damage make it clear just how fragile this machine really is. One of the most tragic diseases to affect decision making is frontotemporal dementia (FTD), a degenerative brain disease that in some cases can cause people to gradually start making terrible decisions as the disease progresses, and which generally comes on much earlier than Alzheimer's disease, often in one's 40s or 50s. Dr. Bruce Miller of the University of California at San Francisco has studied FTD for many years and has amassed a long list of case histories that show just how badly FTD can affect decision making, such as this one:

> Her husband notes increased risktaking over the last 2 years. She has had a number of driving tickets for careless driving, including such things as not stopping at stop signs and backing

down a one-way street instead of going around the other way. These tickets have accumulated over the last 2 years to a degree that they are unable now to get her insurance changed due to the excessive number of driving tickets she has received. By her husband's report she will lose her license if she gets 1 more ticket. Other risk-taking behavior was seen several months ago during a camping trip when she insisted on exploring an out-of-reach waterfall. When her husband explained that it was too dangerous because of unpredictable surf and risk of drowning, she continued to speak about the waterfall and early one morning went to explore it without telling him. She ended up injuring herself pretty badly on the walk. The patient exhibits little insight into these behavioral changes. This patient was subsequently detained by the police after a shoplifting incident and since she was unable to communicate or follow instructions, she ended up on the ground in handcuffs.[3]

The choices that we study in the laboratory are rarely as consequential as these, but they do provide us a window into the brain's machinery for decision making. An example comes from a study that Sabrina Tom in my laboratory performed in collaboration with Craig Fox, a professor in the school of business at UCLA (I described this experiment briefly in chapter 3).[4] We wanted to examine how the brain computes simple choices like the 50/50 chances to win or lose money that I described in the previous section, and to test whether we could see evidence for the predictions of prospect theory in brain activity. We brought 16 volunteers to the MRI scanner and presented them with a large number of gambles where they could win some amount or lose some other amount with a 50/50 chance; the amount that they could win on each gamble varied from $10 to $40, and the amount they could lose varied from $5 to $20. On each trial, they were presented with one of these gambles and asked to tell us whether or not they would take the gamble. It's always a concern whether people will tell the truth in these situations, so we used a sort of economic truth serum. After subjects were finished with the experiment and came out of the MRI scanner, we randomly selected three of the gambles that they had experienced and looked to see whether they had said that they would take the gamble. If they said yes, then we flipped a coin and played the gamble for real money. To make it seem even more real, we gave them $30 in cash a week prior to the scan, and then asked them to bring $60 in cash with them to the scan, which they would be gambling with during the scan. No one actually lost more than the $30 that we had given them (the average subject won $23), but this procedure made the subjects feel as though they were really gambling with their own money and gave us more confidence that we were seeing realistic patterns of decision making in the brain.

When we analyzed the fMRI data, what we found was that there were two areas that showed a striking pattern of response to gains and losses (see figure 7.1). One of them, the *ventromedial prefrontal cortex*, sits in the bottom ("ventral") and middle ("medial") part of the frontal lobe, just behind the bridge of the nose. The other, the *ventral striatum*, is part of a set of brain systems known as the *basal ganglia* that sits deep within the brain. Both of these are areas that neuroimaging has shown play a central role in decision making. In our study, what we saw when we analyzed the trial-by-trial brain activity was that activity in these areas went up as the amount that the subject could win increased, but went *down* as the amount that the subject could lose increased. One of the main goals of the study was to examine the relation between the brain's response to gains and the response to losses, and this is where we were able to establish a direct link between the predictions of prospect theory and the activity of the brain. In particular, prospect theory predicts that losses hurt more than gains feel good. In both of these areas, we saw that the amount that brain activity decreased for bigger losses was steeper than the amount that it increased for bigger gains; we called this "neural loss aversion" by analogy to the concept that Kahneman and Tversky had developed in their studies of behavior.

The final clincher was the relation between brain activity and the behavior of the individual subjects. Some of the subjects in the study were quite averse to losses (only accepting the gamble if they could win more than five times what they could lose),

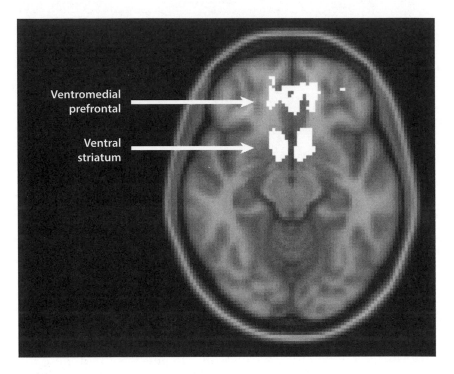

Figure 7.1. Results from our 2007 study of gambling decisions, showing (in white) regions in the ventral striatum and ventromedial prefrontal cortex that responded to increasing gains and decreasing losses.

whereas others just required that the amount they could win was larger than the potential loss by a small fraction. When we compared this to brain activity, we saw that the difference in brain responses to losses versus gains tracked the subjects' choices, such that people who were more averse to losses in their choices also had a larger difference between the response to losses versus gains in their brain. This study provided the first direct evidence from neuroscience in favor of prospect theory.

In science, replication is important before we truly believe a result, and that's particularly true in this study of 16 subjects, given the concerns about small samples that I discussed in chapter 6. Fortunately, the relationship between neural and behavioral loss aversion has since been replicated by a group of Italian researchers.[5] However, another one of the results from our study has not fared as well. Remember that we found several areas in the brain whose activity turned down when the losses got bigger. We were also very interested in trying to find areas where the activity would turn *up* for bigger losses, since many researchers had proposed that thinking about losses should engage brain systems involved in fear or aversion. When we looked for regions whose activity went up with bigger losses, we found nothing. We looked especially hard in the amygdala, which is a region that many researchers had thought should be involved because of its longstanding association with fear and negative emotion (mentioned in chapter 1 and discussed further in chapter 8), but even with our fine-toothed comb we didn't see anything. The subsequent research has shown that we were probably wrong, having missed the signal owing to our lack of statistical power because of the small sample size. The Italian group found that activity in the amygdala did indeed go up when people entertained bigger losses, and other research by Benedetto De Martino and his colleagues at Caltech showed that individuals with damage to the amygdala—the same ones described in chapter 1 who don't experience fear—also didn't show the same level of loss aversion as healthy individuals. I think this is a great example of how individual studies may be right or wrong, but in general science tends toward the truth over time.

Learning What's Good

Several years ago, after almost 20 years of vegetarianism, my wife and I decided to start eating meat again. This left us with a huge landscape of new foods to try, but with little knowledge as to what we would actually like. Should we get the ribeye or the sirloin steak? Other than going by price, we were left asking the butcher, but often that doesn't really help so we ultimately had to take something home and try it out. Usually we liked what we got, but sometimes we didn't, and the next time we made sure not to buy that item again. On the other hand, if we really liked it then we were more likely to buy it again in the future. The ability to learn from trial and error in this way, which we call *reinforcement learning*, is fundamental to survival in the world and has long fascinated scientists. And, as I mentioned in chapter 1, the neurotransmitter dopamine plays a critical role in this kind of learning.

Remember from chapter 1 that dopamine is transmitted broadly across the brain, and that it is released when we experience something that is better than we expected. Through evolution, our brains have learned to use this as a cue that whatever we just did we should probably do again in the future. In fact we now have a detailed understanding of exactly how this works. When we make decisions, there is a competition in our brain to determine which of the many possible actions we will choose, and this competition is centered around a brain circuit that connects the cerebral cortex to a set of deep brain regions called the *basal ganglia*. The strength of the connections between different neurons in the prefrontal cortex and the basal ganglia determines which specific action will be chosen in any circumstance; the stronger the connection, the more likely the action is to win the competition. It so happens that dopamine plays a central role in determining the strength of these connections, in a special version of the kind of Hebbian plasticity that we discussed in chapter 5: if neurons fire together in the presence of dopamine, their connections get stronger, while firing together in the absence of dopamine causes them to get weaker.

Because fMRI is only sensitive to blood flow, rather than to specific neurochemicals, we can't use it to measure dopamine directly. However, there is a lot of evidence that we can see the footsteps of dopamine in the fMRI signal. The first evidence came more than 15 years ago, when researchers found that fMRI signals in the basal ganglia showed a signature of the "reward prediction error" signal that I mentioned above. However, the definitive evidence relating dopamine to fMRI signals in the basal ganglia came from a study in rats by my colleagues Brian Knutson and Karl Deisseroth at Stanford.[6] They used a technique (pioneered in Deisseroth's lab) called *optogenetics*, which allows researchers to control the activity of specific sets of brain cells using light. This technique has revolutionized neuroscience, and will almost certainly win Deisseroth a Nobel Prize. What they did was to use the technique to selectively turn on the activity of the neurons that release dopamine, while the animal was being scanned using fMRI in a tiny version of a human MRI scanner. When these neurons were turned on, releasing dopamine across the brain, they saw increased fMRI signals in several regions including the basal ganglia. This shows definitively that we can see dopamine signals in fMRI data, but remember the lesson of reverse inference: the fact that dopamine causes activity in the basal ganglia doesn't mean that activity in the basal ganglia *must* be due to dopamine, and in fact we are pretty sure that's not the case. This means that neuroimaging signals remain difficult to interpret in terms of their underlying biology.

Our understanding of reinforcement learning has also been greatly aided by computer scientists, who have long been interested in developing machines that can learn through trial and error. Computational models of reinforcement learning are generally based on the idea that we should change our "policy" (that is, the rules that determine how we will behave in certain circumstances) based on how far off our predictions are: bigger surprises should cause a bigger change in the policy. The model computes the value of each possible action in the world and then chooses which action to take at each point based on those values. It then updates its estimates of the values by changing them according to the amount of prediction error; for example, if I choose an action that predicts that I will get one unit of reward, and I end up getting two units of reward, then I will increase the value that I place on that action, which means that I will be more likely to choose it in the future. One of the major successes in the field of cognitive neuroscience, and in neuroeconomics in particular, has been the application of reinforcement learning models from computer science to the analysis of fMRI data, which lets us look for brain systems that respond in the way that is predicted by specific parts of the computational model. For example, we can apply the computational model to individual subjects' behavior as they are learning, and then use the model to estimate what the level of reward prediction error must have been for each choice. We can then look at which brain regions respond in a way that tracks this quantity. This kind of "model-based fMRI" has provided us with a whole new way to better understand what changes in activation actually mean, in terms of the computations that the brain must be performing. It also forms the basis for the computational psychiatry approach that will be described in chapter 8.

Are We Really Two Minds in One?

Many thinkers about human behavior have appealed to a distinction between our rational minds and our "animal spirits." Speaking of the causes of instability in economic markets, the economist John Maynard Keynes said:

> Even apart from the instability due to speculation, there is the instability due to the characteristic of human nature that a large proportion of our positive activities depend on spontaneous optimism rather than on a mathematical expectation, whether moral or hedonistic or economic. Most, probably, of our decisions to do something positive, the full consequences of which will be drawn out over many days to come, can only be taken as the result of animal spirits—of a spontaneous urge to action rather than inaction, and not as the outcome of a weighted average of quantitative benefits multiplied by quantitative probabilities.[7]

More recently, Daniel Kahneman has described this in terms of an impulsive and irrational "system 1" versus a slow and rational "system 2."[8] These stories certainly capture something correctly about human behavior, but I think that rather than reflecting the fact that there are truly two systems in the brain, they primarily reflect that humans are strongly drawn to thinking in dichotomies. The complexity of the brain is enormous, and it seems highly unlikely that any dichotomy will adequately describe it. Having said that, there is at least one such dichotomy that has gained substantial support from neuroscience, which is the distinction between habitual and goal-directed action.

Most of our behavior throughout the day happens without any mental reflection on the choice; we step our foot forward, speak linguistically correct sentences, and drive from the office to our home without having to consciously think about what is going to happen next. This is the power of habit, which is the term that we use to describe actions that occur automatically without cognitive effort or intention. Without habits we would be paralyzed by choice, as noted by William James in the famous chapter on habit in his 1890 *Principles of Psychology*:

> The more of the details of our daily life we can hand over to the effortless custody of automatism, the more our higher powers of mind will be set free for their own proper work. There is no more miserable human being than one in whom nothing is habitual but indecision, and for whom the lighting of every cigar, the drinking of every cup, the time of rising and going to bed every day, and the beginning of every bit of work, are subjects of express volitional deliberation.[9]

An everyday example of the battle between habits and goal-directed actions occurs when we have to detour from a usual route in order to make an unexpected stop (say, at the dry cleaners on the way home from work). Many of us have had the experience of getting home and realizing that we were supposed to have made such a stop, but that we got distracted and habit took over. The nature of a goal-directed behavior is that it requires us to keep an end goal in mind; any distraction that causes the goal to be lost will probably prevent us from doing it. Just as habits are important to prevent paralysis by analysis, goal-directed behavior is equally important to allow us to transcend our usual routines and behave in adaptive ways when the world changes.

The economic model of decision making that we discussed above falls squarely in the domain of goal-directed decision: when we are making a choice, we think about how much we like each of the options and pick the one that we value the most. However, we all know that this fails to describe the process of how many of our decisions are actually made. When I go to the grocery store to buy a tub of yogurt, I don't look at all of the options and think about which one I want the most; I just grab a tub of the same Strauss Dairy plain yogurt that I have bought every week for the past few years. It's only when the world changes, such as when the store has run out of my favorite yogurt, that the goal-directed decision making kicks in. In a sense, this is perfectly rational: Why should I spend precious time thinking about what to choose when I can just do what has always worked for me in the past?

The real problem comes about when those old choices no longer make sense, either because the world has changed or because you have changed. Back when I was a vegetarian, I had a serious M&M habit. Every afternoon I could be found at the candy machine, like a rat pressing a lever to get my daily ration. The main reason for changing my diet was the realization that a high-carb diet was not as good for my health as I thought it was, and over the course of about a year my mind-set about food and health completely changed. What didn't change was my craving for those M&Ms. Sometimes I am able to resist, but it's still the case that if someone sets a bowl of them in front of me, my new mind-set will be a weak match for my habit.

We discussed above how dopamine plays a role in learning about which actions are good and which are not, and this extends to the learning of habits as well. For example, people with Parkinson's disease, who have reduced levels of dopamine because their dopamine-producing cells are slowly dying, do not learn habits in the same way that healthy people do. We found a striking example of this a few years ago, when I collaborated with Daphna Shohamy and Mark Gluck at Rutgers University to study the way that people with Parkinson's disease learn a simple "weather prediction" task.[10] In the task, we showed the subjects sets of cards and asked them to learn by trial and error whether each set of cards predicted rain or sunshine. We made this difficult by giving them noisy feedback—for example, for one set of cards we might tell them that it predicted rain 75% of the time and sunshine 25% of the time. Our previous fMRI research had shown us that this task engages the basal ganglia when subjects are learning by trial and error, but not when they simply try to remember the outcomes without actually making choices.[11] We saw a parallel result when we tested people with Parkinson's disease on the task (without fMRI)—they were very bad at learning which cards predicted rain or sunshine when we asked them to learn by trial and error, but they were able to learn the same information as well as healthy subjects if we asked them to simply watch and learn without responding. This finding fits with a long line of research that has shown that the brain systems involved in learning habits are distinct from those involved in conscious memory for events in the past, and tells us that dopamine is particularly important for the development of habits.

The flipside of this phenomenon can be seen in a bizarre side effect that sometimes occurs with a class of drugs that is used to treat Parkinson's disease and restless legs syndrome. These drugs, called "dopamine agonists" (known by trade names including Requip and Mirapex), trick the dopamine receptors in the brain into thinking that there is dopamine present. The drugs generally help with the symptoms of these diseases, but a small proportion of people taking these drugs started to report a very strange phenomenon: they suddenly became addicted, sometimes to obvious things like gambling, but in other cases to very odd compulsive behaviors. I once had a conversation on a flight to Los Angeles with an accomplished architect who had started taking one of these drugs to treat his Parkinson's disease and suddenly became addicted to gardening; he simply could not stop himself from spending all day in the yard. As soon as he stopped the drug, the insatiable urges to garden went away. This obsession was relatively benign, but many were not so lucky: a petition from the group Public Citizen (calling for a "black box" warning on these drugs) listed a number of cases where the outcomes were much more tragic, including a set of case studies reported by Kevin Klos and colleagues:

- Within six months of starting one of these drugs, a man who had never had any previous interest in pornography started buying pornographic films and having extramarital affairs. He also started gambling and lost hundreds of thousands of dollars, started overeating and gained fifty pounds in six months, and increased his smoking from one to two packs per day. All of these behaviors stopped within one month of stopping the drug, leading his wife to report "I have my old husband back."
- Another man became hypersexual within one month of starting the drug. Whereas they had previously had sex

only once every few months, he suddenly demanded it several times per day. He also tried to pay a friend of his daughter to have sex with him, and requested that his son and daughter-in-law engage in group sex with him. This behavior resolved within a few months of stopping the drug.[12]

These cases show just how powerfully the dopamine system can drive people to engage in their worst impulses. We will discuss the relation between dopamine and habits further in chapter 8 when we discuss its role in drug addiction.

Now versus Later

Many choices that we make come down to a simple decision: Do you want something good now or something better later? When people put away money each month into a retirement account, they are making the decision to forgo spending some money this month so that they will be able to spend it (within compound interest) in the future. When I decide to have a piece of fruit rather than a chocolate flourless cake for dessert, I am trading off the immediate pleasure of the cake against my desire to be healthy in the future, not to mention my desire to not have to buy larger pants. We call these kinds of decisions "intertemporal choices," since they require making a choice between two different goods at two points in time, and they make up many of the most important choices in our lives.

There is an ongoing debate about how the brain accomplishes this kind of choice. One view, which has been championed by the neuroscientist Sam McClure of Arizona State University, is that there are two systems in our brain, much along the lines of Kahneman's system 1/system 2 distinction: an impatient emotionally based system that is constantly screaming "now!" and a more patient rationally driven system that nudges us to calmly consider the benefits of waiting. In McClure's first fMRI study of intertemporal choice, he presented subjects with choices between a smaller amount of money available sooner (for example, $5 today) and a larger amount available later (say, $20 in six months) while he measured brain function using fMRI. When he compared activity between trials where the individual chose the immediate reward and those where the delayed reward was chosen, he saw that immediate choices were associated with more activity in a set of brain regions that by now will have become familiar: the ventral striatum and ventromedial prefrontal cortex, both strongly associated with reward processing. Conversely, choices of the delayed response were associated with greater activity in areas including the dorsolateral prefrontal cortex, which is thought to support top-down executive control over our decisions. The researchers concluded: "In economics, intertemporal choice has long been recognized as a domain in which, 'the passions' can have large sway in affecting our choices. Our findings lend support to this intuition."[13]

The intuition of passion battling it out with reason was not shared by Paul Glimcher. A professor at New York University, Paul was one of the early pioneers of the field of neuroeconomics, having done seminal work showing that the responses of single neurons in a monkey's brain are related to the value that the monkey places on stimuli in the world. Along with postdoctoral fellow Joe Kable, Glimcher set out to test a different way of thinking about intertemporal choice that was flavored by his years studying single neurons. One of the striking findings from the monkey studies is that one can often establish a close relationship between the activity of a single neuron and the behavior of an animal. This is known as a "psychometric-neurometric comparison"—with "psychometric" referring to the animal's decisions and "neurometric" referring to the activity of neurons. Previous research had shown that the relationship between these two can be very orderly, though those results came from studies of simple visual stimuli such as moving dots. However, if decisions are based on an underlying computation of value, then Kable and Glimcher expected that there should be brain areas in people where there is an orderly relationship between the person's subjective value of the option and his or her brain's response to this option—a psychometric-neurometric match. To measure this, they used an intertemporal choice task very similar to the one used by McClure, where people made choices between a constant immediate amount ($20 today) and a range of larger delayed amounts at various delays.[14] They took a different approach to analyzing their data, which was to look for brain regions where there was a direct relationship between the person's subjective value for the delayed option (which they inferred from the subject's choices) and the brain's response across trials. They found that a psychometric-neurometric match was indeed present in some of the same areas that McClure's work had labeled as "impatient": the ventral striatum and ventromedial prefrontal cortex. In a knockout punch to McClure's hypothesis of dueling systems, they also analyzed their data using the same model that McClure had used and showed that their simpler model actually described the data better than McClure's model involving separate impatient and patient brain systems. These results convinced many researchers in the field that one doesn't need to appeal to a battle between passion and reason in order to explain how we sometimes are willing to wait for a bigger reward and sometimes simply want to take what we can get right away.

Another striking finding from the Kable and Glimcher study was that among their 10 subjects there were huge differences in the degree to which they discounted future rewards; some were incredibly patient while others wanted it all now. We all know intuitively that some people are better at waiting than others, and this ability seems to have an important impact on a person's ability to succeed in life. One of the best-known demonstrations of this comes from the "marshmallow study" by Walter Mischel and colleagues. In this test, children are taken into a room and shown two treats (such as a marshmallow and a cookie) by a researcher, and told that if they wait for the tester to return, they can have both treats, but if they decide that they don't want to wait they can ring a bell and have just one of the treats immediately. More than 500 children at the Bing Nursery School at Stanford were tested by Mischel and his colleagues in the late 1960s and early 1970s, who then followed them up over time to see how the ability to delay gratification was related to later outcomes in life. What they found was striking: the children who were better able to delay gratification were more academically successful (performing better on the SAT years later) and were also described by their parents as being more academically and socially adept.[15]

B. J. Casey is one of the world's leading cognitive neuroscientists, and in 2011 she and her colleagues (including Walter Mischel) reported a study that gave us new insights into the differences in brain function between Mischel's patient and impatient kids. They followed up with a group of Mischel's children, roughly 40 years after they had been first tested, and were able to convince 27 of them to return to the lab for MRI scanning. They used a task that Casey has developed over many years of research, called a "go/no-go" task, in which subjects are shown some stimuli and have to respond to most of them but occasionally withhold their response (the same task used by Kent Kiehl in the neuroprediction study mentioned in chapter 6); in this case, the stimuli were happy or fearful faces, and the subject had to respond to one emotion but withhold his or her response to the other. We know from previous research from my lab and many others that when people stop themselves from making a response, a network of brain regions is activated that centers on a region in the right prefrontal cortex called the inferior frontal gyrus.[16] When Casey compared her subjects who had been good delayers to those who had been poor delayers, she saw that the poor delayers were worse at stopping themselves (making more errors when they were supposed to withhold their response), and that their inferior frontal gyrus was less active than the good delayers. This is an exciting finding that shows the echoes of early childhood behavior decades down the road, but it's worth noting that the ultimate sample was very small (with only 15 high-delaying individuals and 11 low-delayers), so we need to interpret the results with caution and wait for other larger studies to replicate this finding.

The Advent of "Consumer Neuroscience"

It was really only a matter of time until neuroeconomics caught the attention of people whose job it is to more effectively sell us things. With billions of dollars at stake, even tiny improvements in the market share of a product can translate into huge sums of money. The potential of "neuromarketing" (or, as its advocates now call it, "consumer neuroscience") was first made clear in a 2004 study by Sam McClure and Read Montague, which has become widely known as the "Coke/Pepsi study."[17] McClure was interested in understanding how brand information changes the brain's response to a food, and cola drinks are a particularly interesting example because it has long been known that people's stated brand preferences often don't match their choices in a blind taste test. To test this, McClure first asked people which brand they preferred, and then also gave them a blind taste test—indeed, there was very little relation between the brand that people said they preferred and what they chose in the blind taste test. He then used a custom device to deliver soda to people while they were being scanned using fMRI, which let him measure the brain's response to these drinks either with or without knowledge of the brand. He focused on the ventromedial prefrontal cortex, which you will remember from earlier as one of the areas that shows a strong response to rewards. When people were delivered drinks without any brand information, the response in this area turned up in direct relation with how much the person said he or she liked the drink.

What about the effect of knowing the brand? It turned out that the Coke brand was stronger than the Pepsi brand, which could be seen both in behavior and in the brain. In a taste test where people were given the chance to drink Coke versus an unlabeled drink that they were told could be either Coke or Pepsi, people were more likely to choose Coke; however, in the same experiment done with Pepsi, there was no effect of the brand on people's choices. fMRI showed that there was more activity in several brain areas when people received branded Coke compared to Coke without the brand information, including the dorsolateral prefrontal cortex, which seems to play a role in exerting top-down effects on our fundamental desires; here, too, no differences were found for Pepsi. This study provided the first suggestion that it might be possible to use neuroimaging to identify the effects of marketing information.

Several years after the Coke/Pepsi study, one of the first procla-
mations of the power of neuromarketing appeared in one of my
favorite sources for overblown junk science: the *New York Times*
op-ed pages. A team of cognitive neuroscience researchers from
UCLA (including Marco Iacoboni, who was also behind the "Your
Brain on Hillary" debacle discussed in chapter 1) used fMRI with
television ads presented during the 2006 Super Bowl in order to
see how five viewers' brains would respond to them. Later that
evening, Iacoboni had already picked the winning and losing ads
based on how much activation they caused in areas thought to be
related to emotion, empathy, and reward:

> Who won the Super Bowl ads competition? If a good indicator
> of a successful ad is activity in brain areas concerned with
> reward and empathy, two winners seem to be the "I am going
> to Disney" ad and the Bud "office" ad. In contrast, two big
> floppers seem to be the Bud "secret fridge" ad and the Aleve
> ad. What is quite surprising, is the strong disconnect that
> can be seen between what people say and what their brain
> activity seem to suggest. In some cases, people singled out ads
> that elicited very little brain responses in emotional, reward-
> related, and empathy-related areas.

> Among the ads that seem relatively successful, I want to single
> out the Michelob ad. Above is a picture showing the brain
> activation associated with the ad [*not shown here*]. What is
> interesting is the strong response—indicated by the arrow—
> in "mirror neuron" areas, premotor areas active when you
> make an action and when you see somebody else making the
> same action. The activity in these areas may represent some
> form of empathic response. Or, given that these areas are also
> premotor areas for mouth movements, it may represent the
> simulated action of drinking a beer elicited in viewers by the
> ad. Whatever it is, it seems a good brain response to the ad.[18]

If this book has done its job so far then your hackles should be
on end, as this is a prime example of reverse inference gone wild.
The second paragraph of the quote makes that abundantly clear:
He tells us that there is activity in the "mirror neuron" areas

of the brain, which could have to do with empathy, but could also have to do with simulated mouth movements. In this case interpreting fMRI results seems to be more akin to a Rorschach test (identifying what the interpreter already believes) than to actual science.

Several years later, the *New York Times* op-ed page once again published overblown neuromarketing claims, this time by self-proclaimed "neuromarketer" Martin Lindstrom in an article titled "You Love Your iPhone. Literally."

> Earlier this year, I carried out an fMRI experiment to find out whether iPhones were really, truly addictive, no less so than alcohol, cocaine, shopping or video games. In conjunction with the San Diego-based firm MindSign Neuromarketing, I enlisted eight men and eight women between the ages of 18 and 25. Our 16 subjects were exposed separately to audio and to video of a ringing and vibrating iPhone ... most striking of all was the flurry of activation in the insular cortex of the brain, which is associated with feelings of love and compassion. The subjects' brains responded to the sound of their phones as they would respond to the presence or proximity of a girlfriend, boyfriend or family member. In short, the subjects didn't demonstrate the classic brain-based signs of addiction. Instead, they loved their iPhones.[19]

If Lindstrom is correct, then activation in the insular cortex of the brain provides a neuroscientific "love meter." Unfortunately this is simply incorrect, as I and 44 colleagues pointed out in a letter to the editor of the *New York Times*:

> "You Love Your iPhone. Literally," by Martin Lindstrom (Op-Ed, Oct. 1), purports to show, using brain imaging, that our attachment to digital devices reflects not addiction but instead the same kind of emotion that we feel for human loved ones. However, the evidence the writer presents does not show this. The brain region that he points to as being "associated with feelings of love and compassion" (the insular cortex) is active in as many as one-third of all brain imaging studies. Further, in studies of decision making the insular cortex is more often

associated with negative than positive emotions. The kind of reasoning that Mr. Lindstrom uses is well known to be flawed, because there is rarely a one-to-one mapping between any brain region and a single mental state; insular cortex activity could reflect one or more of several psychological processes. We find it surprising that The Times would publish claims like this that lack scientific validity.[20]

Marketers continue to regularly make overly strong claims about the power of neuroscience methods. To combat this, Joe Devlin of University College London came up with a list of warning signs for what he calls "neuromarketing snake oil." The first warning sign, "Beware claims of mind reading," is in line with the point made throughout this book that there is no simple one-to-one mapping between psychological states and brain activity. Another warning sign is "Beware proprietary data analysis techniques." While it might seem that neuromarketers could use their substantial financial resources to come up with better ways to analyze data, in reality such claims are usually a cover for untested methods with unknown reliability. Others include being on the lookout for "neuro-sophisms and neuro-myths" as well as for "quack neuroscientists" with seemingly impressive titles like "Chief Neuroscientist." More generally, it's important when reading about scientific research of any kind to be sensitive to overselling and oversimplification—if it seems like you are being given a sales pitch, then you probably are, and that's usually not the sign of solid research.

In the wake of these ham-handed early efforts to establish neuroscience as a tool for marketing, a growing group of researchers is taking a more careful and reasoned approach to investigate whether neuroscience is actually worthwhile for understanding consumer behavior. One of the leaders of this group has been Vinod Venkatraman, a marketing researcher at Temple University who has undertaken one of the most systematic examinations of the effectiveness of various neuroscience tools for marketing purposes.[21] Alongside standard marketing questionnaires about how much people liked each of several products, he and his colleagues tested a number of different methods, including fMRI and EEG along with other measures meant to measure physiological responses (such as heart rate measurement) and psychological measures meant to measure implicit attitudes. They measured the response to each of a large number of ads, and then used data on real-world advertising effectiveness to determine which methods were best at predicting the effectiveness of the ad campaigns. Their results showed that fMRI (in particular, response in the ventral striatum to the ads) was far more effective than the other methods at predicting ad effectiveness, improving the accuracy of the predictions by more than 50%. The improved effectiveness of fMRI may reflect the fact that the response of the ventral striatum is either more sensitive to the features of products that drive people to want them, and/or that these responses are less sensitive to other factors that can cause psychological measures to be unreliable. This kind of careful work shows that there is indeed promise in the application of neuroscience to marketing questions.

Reading the Collective Mind

The ability to predict the behavior of an entire population based on the brain activity of a few people is the holy grail of a number of fields, such as consumer neuroscience and neuropolitics, and new research has started to suggest that it might actually work. One of the leaders in this field is Emily Falk, who is now a professor of communication at the University of Pennsylvania. She has pioneered the idea of the "neural focus group," in which fMRI is used to predict how people will respond to an advertisement at the population level. In her first study to examine the effectiveness of the neural focus group, she used fMRI to predict responses to an antismoking campaign for the US National Cancer Institute's quitline (1-800-QUIT-NOW).[22] Subjects viewed a set of television advertisements during scanning, which were taken from three different advertising campaigns. The subjects also answered a questionnaire about how effective they thought the ads were, which is what the standard focus group would do. Falk examined brain activity evoked by each ad within the ventromedial prefrontal cortex; this is the same area that we previously discussed in the context of decision making, whose

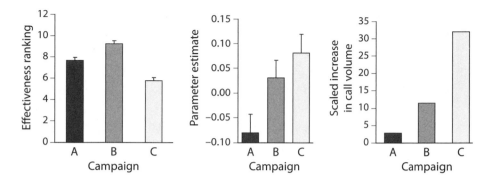

Figure 7.2. Results from Emily Falk's neural focus group study. The *left panel* shows how effective the subjects rated each ad; they thought that the ads from campaign B were the most effective. The *middle panel* shows the average brain response to ads from each campaign; the response to ads from campaign C was largest. The *right panel* shows the actual call volume to the quitline after each campaign, which followed the brain response rather than the subjects' survey responses. From Emily B. Falk, Elliot T. Berkman, and Matthew D. Lieberman, "From Neural Responses to Population Behavior," *Psychological Science* 23, no. 5 (2012): 439–45, copyright ©2012 by Emily Falk. Reprinted by permission of SAGE Publications, Inc.

activity generally increases in response to stimuli that individuals value positively. She and her colleagues then pitted the results from this "neural focus group" against the standard survey group by comparing the results from both to a very different real-world outcome: the volume of phone calls to the quitline in the month following the introduction of each ad campaign. The results (shown in figure 7.2) were striking: the brain responses of the 30 subjects correctly predicted which ad campaign would be most effective, while the standard focus group responses did not.

The idea of the neural focus group inspired my colleague Brian Knutson and his student Alex Genevsky to ask an even harder question. They were interested in understanding how people make decisions on the microlending website Kiva, in which individuals from low-income countries post requests for funding of their small businesses. One of the major benefits of analyzing the Kiva site is that it allows researchers to obtain extensive data automatically, providing the ability to assess a large number of lending decisions; Genevsky and Knutson used this to obtain data on almost 14,000 decisions, half of which were decisions to lend and the other half were decisions not to lend.[23]

They first asked whether they could predict lending decisions based on the features of the photo of the individual requesting funding or the description of the loan request. Because it would have been very difficult for one person to sit down and rate the features of so many photos, they instead used a tool that has become a mainstay of psychological research: Amazon's Mechanical Turk (known among researchers as "MTurk"). MTurk is an online marketplace for workers to complete web-based tasks. Individuals (known as "Turkers") are able to choose among a large number of tasks, each of which offers to pay a certain amount of money for a certain amount of work. In this case, each Turker was presented with one of the photos and asked to rate its visual clarity as well as what emotion the person was displaying. Using these data alongside the data obtained from the Kiva site, Genevsky and Knutson were able to show that one can successfully predict which loan requests will be successful based on several features of the photo, including the sex of the requester (women were more successful than men), the clarity of the image, and the amount of positive emotion portrayed by the requester in the photo. The accuracy was not earth-shattering—knowing just the amount of positive emotion in the image, they could predict about 17% of the variability in Internet lending rates, which is better than guessing but far from perfect.

They next asked whether the brain responses of a neural focus group could help improve prediction. They chose 80 requests that had the most extreme ratings of positive or negative emotion from the large set rated by the Turkers, and then presented these to subjects during fMRI scanning while they made lending decisions about each request. When they compared activity between loans that subjects funded and those that they didn't, there was activity in the ventral striatum and ventromedial prefrontal cortex, two regions that we have seen repeatedly to relate to subjective preferences. They quantified the response in the ventral striatum, which, as we have seen, is often associated with reward processing. Using just the fMRI data (see figure 7.3), they were able to account for about 6% of the variability in lending rates; note that while this is better than nothing, it's substantially less than one could predict using just

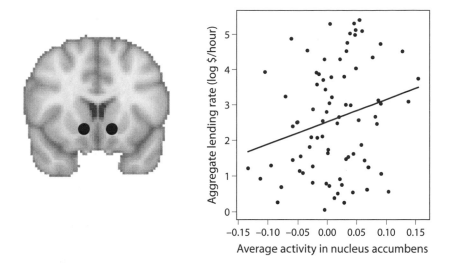

Figure 7.3. Results from the neural focus group study by Alex Genevsky and Brian Knutson. The *left panel* shows the location (in black) of their regions of interest within a part of the ventral striatum called the nucleus accumbens. The *right panel* shows the relationship between activity in the region of interest that was evoked by each request, and the amount of lending that each request received on Kiva, with the best fitting line showing the average relationship between the two. Generated using data generously provided by Alex Genevsky and Brian Knutson.

the photo ratings. Putting together the photo ratings and the fMRI data, they were able to reach a prediction accuracy of 20% of the variance, showing that combining the photo ratings and fMRI data was better than either alone.

We don't know whether marketing companies are currently using neural focus groups, but it's a pretty good bet that they are. Even though the levels of prediction accuracy are often only slightly better than guessing, these small improvements could still lead to big financial gains if they improve sales even just a tiny bit. We still need more data in order to understand just how generalizable these results are—for example, do they only work with ads involving photos of people?—but it's almost certain that these questions are currently being asked by consumer neuroscientists and will play an increasingly important role in advertising, politics, and other areas where public opinion is important for decision making.

The degree of interest in neuroscience within the business world is evident in the fact that many major business schools have hired faculty who engage in neuroscience research, including Harvard, Stanford, and the University of Chicago. The Wharton School of Business at the University of Pennsylvania has started the Wharton Initiative in Neuroscience, which is assembling a set of researchers dedicated to "building better business through brain science." To run this initiative they tapped Michael Platt, a neuroscientist who had spent most of his career studying decision making in monkeys and who had collaborated with Paul Glimcher on some of the earliest studies of value-based decision making in monkeys. The interest is also evident in the fact that Nielsen, the premier market research company in the United States, has established a consumer neuroscience team of almost 20 researchers, whose job it is to figure out how to use neuroscience to improve market research. The interplay between neuroscience and business shows no signs of stopping.

CHAPTER 8

IS MENTAL ILLNESS JUST A BRAIN DISEASE?

The first time I got into an MRI scanner, I had a severe anxiety attack. The scanner had a panic button that I could press to alert the operator that I needed help, but this was such a major-league freak-out that I ended up crawling out of the scanner before they could get into the room and wheel me out gracefully. I've now been in MRI scanners more than 100 times (as you learned in chapter 5), and I've gotten pretty comfortable with it, but I sometimes still have flashbacks to that first scan whenever the bed starts sliding into the scanner. That panic attack in the scanner was not my first encounter with anxiety—in fact, at that point I had been dealing with it for a number of years. The first anxiety attack that I can remember happened just after I had started as a graduate student. I raised my hand to ask a question in a seminar, and suddenly was overcome with all of the telltale signs of panic: my heart was racing, I wasn't sure if I could breathe, and my face suddenly felt very hot. For an aspiring academic, this is not a good start to one's career. And throughout the next decade, every encounter with public speaking became an occasion for fear and loathing. It was only through many sessions of cognitive behavioral therapy that I was able to finally kick the fear of speaking in public, or at least tamp it down to a level that doesn't make me miserable.

Anxiety is just one of a set of mental health problems that takes an enormous toll on modern society. It is estimated that roughly 1 in 5 people in the United States experiences some kind of mental illness every year, and 1 in 25 suffers from a severe mental illness that affects his or her ability to function in society. The costs to society are enormous, in dollars and lives; major mental illness costs the United States almost $200 million in lost productivity every year, and suicide is the second most common cause of death for adolescents and young adults. And while death and disability from many other diseases are on the decline, disability from mental disorders is actually rising, despite the massive investment of research dollars by countries around the world.

Until relatively recently, mental health problems were not viewed as diseases of the same sort as cancer or diabetes. Instead, they were seen as manifestations of personal weakness or of spiritual problems such as demonic possession, witchcraft, or undue influence of the moon (which is the source of the term "lunatic," after Luna, the Roman goddess of the moon). Sigmund Freud tried to explain severe mental disorders in terms of sexual urges and our need to repress them; for example, in describing the case of a German judge who became schizophrenic, he said that "we may regard the phase of violent hallucinations as a struggle between repression and an attempt at recovery by bringing the libido back again on to its objects."[1] It wasn't until the twentieth century that physicians and scientists began to view mental illnesses as brain diseases. A major factor in this change was the introduction of drug treatments for major mental illnesses in the 1950s, which convinced many physicians and scientists that these diseases must reflect a "chemical imbalance" that is corrected by the drug. Another important contributor was increasing knowledge about the genetics of mental illness.

What Genetics Has Taught Us about Mental Illness

Genetics plays an important role in all of the major mental disorders, which shows that they must have a biological underpinning. One way to see the role of genetics is to compare identical twins (who share almost exactly the same genome) with fraternal twins (who on average share half of their genome with one another). If one identical twin is diagnosed with schizophrenia, then the other twin has a 30%–50% chance of being diagnosed; for comparison, if they have a fraternal twin with schizophrenia then the rate is 5%–10%, whereas the rate for the entire population is about 1%. Another way to describe the influence of genetics is *heritability*, which is a notoriously difficult concept that describes the degree to which differences across people in some characteristic (such as a diagnosis of schizophrenia) are due to genetic differences versus other differences (such as experience or environment).[2] If a disease has a heritability of 100%, that means that any differences between people in the presence of the disease are due entirely to differences in their genetics; but, importantly, this doesn't mean that the effects of the gene are necessarily inevitable. The best example of this is a rare disease called *phenylketonuria* (PKU), in which a person is born without the ability to process a specific amino acid called phenylalanine. If untreated, the disease results in brain damage and intellectual disability. PKU is 100% heritable, meaning that if both parents carry the gene then the child is certain to have the disease. However, despite the certain inheritance of the disorder, the consequences of the disease can be reduced or eliminated simply by minimizing the amount of phenylalanine in the diet; if you have ever noticed the warning "Phenylketonurics: Contains phenylalanine" on a can of diet soda, this is why.

PKU is due to mutation in a single gene; for this reason it's called a "simple" genetic disorder. Another tragic example of a simple genetic disorder of the brain is Huntington's disease, which causes uncontrollable movements of the body as well as psychiatric symptoms such as mood problems and psychosis. Huntington's disease is caused by a mutation in a single gene, and its inheritance is described according to the rules that Gregor Mendel first discovered in his studies of plants; if a parent has the Huntington's disease gene, then the children have a 50/50 chance of inheriting the disorder, in which case they are guaranteed to develop the disease later in life (usually by about 50 years of age). Unlike Huntington's disease, most major mental disorders are called "complex" genetic disorders because, while they all have a substantial degree of heritability, they are not inherited in a way that would suggest that they are caused by a single gene.

The powerful genetic technologies that were the fruits of the Human Genome Project have enabled researchers to begin to clarify the genetic underpinnings of complex disorders such as schizophrenia, depression, and autism, providing a more direct biological explanation for the heritability of these disorders. An important tool that has enabled the search for genes involved in complex brain disorders is the *genome-wide association study*, or GWAS. GWAS relies on the fact that there are a relatively small number of places among the roughly three billion locations in the human genome where people differ from one another very often—where "relatively small" means something like a few million places, and "very often" means that at least about one percent of people have a different letter in their genome (or *variant*) at that location. The rest of the genome is identical across nearly all humans. To perform a GWAS, researchers collect genetic material from a large number of individuals (from either blood or saliva), and then identify which version of each of those million or so "common variants" is carried by each individual. They then compare all of these variants between a set of people who have the disorder of interest (who are known as "cases") and a similar set of people who don't have the disorder (known as "controls"), to see whether there are any variants that are more or less common in cases versus controls.

As researchers began to use GWAS to investigate diseases such as schizophrenia, it became clear that there were many different genes that appeared to play a role in these diseases, but each individual gene appeared to play a very small role; none of them accounts for more than about one percent of the differences in the presence or absence of the disease. That is, there is no "schizophrenia gene." However, these studies have provided important new insights into the biology of schizophrenia. Surprisingly, the strongest differences in the genomes of schizophrenic individuals are not found in genes with any obvious relation to brain function; instead, they are found in genes related to the function of the immune system, specifically the part that helps the body decide which cells are foreign and which are not (known as the *major histocompatibility complex*). Detailed studies of the genomes of schizophrenic individuals allowed Steven McCarroll and his colleagues at Harvard Medical School to identify one specific gene known as *C4* that differed on average between healthy and schizophrenic people, with the schizophrenics being more likely to have a version of the gene that is more active, meaning that it generates more of its particular protein.[3] They then used studies of mice to understand the role of C4 protein in the brain, and found that it plays a role in the elimination of synapses between neurons, which happens during early brain development; greater C4 activity during early brain development could in theory lead to later brain dysfunction by causing too much elimination of synapses. This work is a shining example of how genetics can lead to a deeper understanding of the biology of mental illnesses, though it's worth remembering that it's still only a small part of the explanation for the disease, because most people with schizophrenia do not have a disordered version of the *C4* gene, so other genes must also be playing a role, given the high heritability of schizophrenia.

What can we conclude from genetic differences in case-control studies? Think back to our discussion of reverse inference. There we saw that the presence of activation in the context of some particular psychological function (such as activation of the amygdala when a person experiences fear) does not tell us that the region is a "fear" region; we have to know what other psychological functions also activate the region. Similarly, finding an association between a genetic variant and a disease doesn't tell us that the gene is specific for that disease; just because a gene is more common in people diagnosed with depression doesn't make it a "depression gene." To ask that question, we need to ask whether the gene is specifically related to the disease, or whether the same genetic variant is also related to other brain disorders. As genetics researchers have begun to ask those questions, the answer has become clear: there is very strong overlap in the genetics of seemingly distinct mental health disorders. In particular, there is a large amount of overlap in the genetics of severe mental illnesses such as schizophrenia, bipolar disorder, and major depression, with weaker overlaps between these and disorders such as ADHD, obsessive-compulsive disorder, and autism. There are many ways to interpret this, but one of the simplest is that a diagnosis of a particular mental disorder, while it might be useful for the practicing psychiatrist, may not be a very useful description at the biological level. As neuroimaging researchers have examined psychiatric disorders, a very similar story has emerged.

Imaging Mental Illness

If mental illnesses are truly brain diseases then we should see evidence of them using brain imaging, and it should come as no surprise that a large number of studies have examined this question. The majority of these studies have examined the structure of the brain, focusing on measurements such as the thickness or density of the gray matter. This is a relatively easy measure to obtain, requiring only a few MRI scans taking 20–30 minutes.

Many of the individual studies published on this topic have demonstrated differences between patients and control subjects, but interpreting these differences is much more challenging because many of the studies have been relatively small, owing in part to the fact that collection of MRI data is much more difficult than taking a cheek swab for a genetics study. However, it is possible to combine the data across many studies, using a technique called a "meta-analysis," which gives us more confidence in the results. The largest meta-analysis of structural MRI in mental illness to date was led by Amit Etkin, a psychiatrist at Stanford. Etkin and his group combined data from almost 200 published case-control studies examining a wide range of mental disorders: schizophrenia, bipolar disorder, depression, addiction, obsessive-compulsive disorder, and anxiety.[4] Their analysis focused on studies that measured differences in the amount of gray matter across the entire brain.

The findings of the meta-analysis were impressive. Across all of the different disorders, there was evidence for less gray matter in people with mental disorders in three regions of the brain: the anterior cingulate cortex, and the left and right anterior insulae (see color plate 10). These are regions that are strongly connected to one another, and a large body of research has suggested that this network is particularly important for *executive function*—the ability to control one's behavior and act in a goal-directed manner. Indeed, when Etkin's group looked at another data set of healthy people they found that gray matter density in this network was correlated with a measure of executive function derived from several different psychological tests. There were some differences between different patient groups; for example, people with "internalizing disorders" (including depression, anxiety disorder, and obsessive-compulsive disorder) showed reduced gray matter in the hippocampus and amygdala, and people with psychotic disorders showed greater overall gray matter loss than nonpsychotic individuals. However, the relative overlap between these seemingly different disorders suggested that their biology may be much more similar than their different symptoms would suggest—a message strikingly similar to the genetics findings that we discussed above, where it was seen that different psychiatric diagnoses can have overlapping genetic causes.

One difficulty with interpreting studies of gray matter density or thickness is that there is not a direct relationship between the amount of gray matter and its function. It might seem that a thicker cortex would be better, but this is not always the case—in fact, the thickness of the cortex decreases from the day we are born through old age, such that a babbling 18-month-old child has thicker gray matter than a 30-year-old math genius. Sometimes thinner cortex is bad, as in Alzheimer's disease, where the cortex becomes overly thin. However, sometimes thicker cortex is bad; for example, many studies have reported that autistic individuals have thicker cortex in some parts of the brain compared to healthy controls. Structural MRI studies can thus tell us whether the brain differs in these disorders, but can't really tell us what it means.

Psychiatric researchers have also used fMRI to look for differences in brain function between healthy people and people with different mental health disorders. Here too there are many individual studies that report differences in brain function between cases and controls, but the studies are often very small, so once again meta-analysis can provide us a consensus view. Emma Sprooten and her colleagues at the Mount Sinai School of Medicine performed such a meta-analysis using data from 537 published fMRI studies of schizophrenia, bipolar disorder, major depression, anxiety disorders, and obsessive-compulsive disorder.[5] The results of this analysis were consistent with those from the structural MRI meta-analysis by Etkin and his colleagues: when Sprooten looked at analyses that examined the whole brain (which are less biased than analyses focused on specific regions), she found that there were consistent differences in functional activity in a number of brain regions between healthy people and people with mental illness, but these brain differences were largely the same across all of the disorders. That is, people with depression, bipolar disorder, and schizophrenia show differences in brain function compared to healthy people, but the differences are largely similar across all of those disorders. As I will discuss later, results like these have led some to question whether our approach to studying mental disorders needs to be completely rethought.

Challenges of Imaging Mental Illness

One problem with interpretation of many neuroimaging studies of mental illness is that it's difficult to disentangle the disease process itself from the effects of the disease, including the drugs used to treat the disease. The best example of this effect is seen in schizophrenic individuals who have been treated with so-called "typical" antipsychotic drugs (such as haloperidol), which block a particular kind of dopamine receptor (known as D2 receptors). Studies that have followed individuals over time after they started taking these drugs have found that in most parts of the brain there is a decrease in gray matter, but in the basal ganglia (which have a large number of dopamine D2 receptors) there is actually an increase in volume. In addition, the lifestyles of mentally ill people are often not conducive to healthy eating, exercise, or social engagement, all of which can have an impact on the brain. Any study that compares healthy individuals with people with a long history of mental illness will be confounded by the effects of medication and other consequences of the disease rather than the disease process itself. One way to address this is to study people when they first become ill, which lets us see the effect of the disease without the confounding effects of years of drug exposure and (in some cases) social isolation. Getting an unmedicated actively psychotic person into an MRI scanner is very difficult, but some studies have succeeded in doing this, and they have shown us that there are indeed differences in brain function between healthy people and people diagnosed with major mental disorders. For example, research by Jong Yoon at Stanford and Cameron Carter at the University of California at Davis has shown that even in schizophrenic individuals having their first psychotic episode and who have never been exposed to antipsychotic drugs, the brain response in regions related to the dopamine system is abnormal.[6]

Another approach that can sidestep the effects of drugs is to study unaffected first-degree relatives (siblings, parents, and children) of mentally ill people. First-degree relatives should, on average, share about half of their genome, and the heritability of mental illness tells us that these individuals should thus share at least some of the genetic vulnerability that contributes to the disease in their affected relatives. Thus, by studying unaffected relatives we can better understand the brain differences that are associated with risk for the disease. Studies of first-degree relatives have indeed shown that they exhibit some of the same brain abnormalities as mentally ill people. For example, research by the late Larry Seidman and his colleagues from Harvard Medical School examined the size of the hippocampus, a brain area involved in forming new memories, in people with schizophrenia and their unaffected relatives. They found that both the schizophrenics and their relatives (especially those from families with multiple people with the disease) had smaller hippocampi, as well as having worse memory compared to healthy controls from families without any schizophrenic relatives. This provides perhaps even stronger evidence that mental illness really is a brain disease, because we can see the lingering traces of genetic risk for the illness even in people who have not actually developed the disease. Neuroimaging has been uniquely important to obtain these insights; while in theory it would be possible to examine the brains of people after their death and compare the size of brain areas across people, in practice such studies are almost impossible to perform, whereas neuroimaging data can be collected fairly easily.

Rethinking Mental Disorders

Results like those outlined above have left researchers who study the biology of psychiatric disorders increasingly frustrated, and much of the blame for this frustration has been laid on a single book: the *Diagnostic and Statistical Manual of Mental Disorders*, better known as the *DSM*. This is the book that defines how psychiatrists diagnose particular disorders. For example, in order to diagnose someone with panic disorder, they must have recurrent panic attacks that include at least four of the following symptoms:

- palpitations, pounding heart, or accelerated heart rate
- sweating
- trembling or shaking
- sensations of shortness of breath or smothering
- feelings of choking
- chest pain or discomfort
- nausea or abdominal distress
- feeling dizzy, unsteady, light-headed, or faint
- chills or heat sensations
- paresthesias (numbness or tingling sensations)
- derealization (feelings of unreality) or depersonalization (being detached from oneself)
- fear of losing control or "going crazy"
- fear of dying[7]

Other disorders are similarly defined in terms of checklists, requiring that the individual experiences a minimum number of the symptoms in order to receive the diagnosis. This smorgasbord approach means that two people can have completely different sets of symptoms—for example, one with palpitations, sweating, shaking, and choking sensations, and another with nausea, dizziness, chills, and tingling—but both would be diagnosed with the same disorder.

There is something clearly very different about this approach compared with how most other nonpsychiatric diseases are diagnosed. For example, the symptoms of diabetes can include extreme thirst or hunger, frequent urination, and fatigue. However, physicians would never diagnose a person with diabetes simply because he or she reports some number of these symptoms. Instead, they draw blood and measure the level of specific chemicals in the blood, which we refer to as "biomarkers," such as glucose or hemoglobin A1C. Similarly, an emergency room physician would never diagnose someone as having a heart attack simply based on a symptom of chest pain. The genomics revolution has taken this even further for cancer diagnosis. Previously, cancer would have been characterized in terms of where it occurred ("brain cancer") or the kind of cells that are involved ("glioblastoma"), but research has shown that this is not very useful in determining how to treat the cancer effectively. With the development of the ability to determine the exact genetic mutations within a tumor, physicians can now use this knowledge to determine the appropriate care; for example, there are four different types of glioblastoma (an aggressive form of brain cancer) that are characterized by different patterns of genetic abnormalities, and which differ in their response to treatment. It is this kind of "precision medicine"—using the knowledge of biology to drive personalized treatments—that inspired one psychiatrist to try to change the way that psychiatric disorders are defined and studied.

Tom Insel was not the person who many would have expected to lead a revolution in psychiatry. He was trained as a psychiatrist, but then spent many years as a basic neuroscience researcher. He was best known for his pioneering research into the sex lives of prairie voles (a small rodent known for being especially social), which showed that specific neurochemical systems were responsible for making these animals monogamous. However, in 2002 Insel was appointed as the director of the National Institute for Mental Health (NIMH), which provides funding for the majority of research into mental health disorders in the United States. This put him into a position to set the research agenda for mental health, and he took full advantage of his bully pulpit. He was particularly struck by the fact that while death and disability from most diseases such as cancer and heart disease have decreased over recent decades, the toll of psychiatric disorders has increased over the same time. He realized that the field of psychiatry had failed to deliver on the promise of making people healthier, and as he said in a 2013 blog post: "Patients with mental disorders deserve better."[8]

The solution that Insel conceived along with psychologist Bruce Cuthbert is known as Research Domain Criteria, or RDoC. The basic idea is that if mental health disorders are brain disorders, then we need to understand the brain systems that go awry in mental illness before we can understand how to diagnose and treat them properly. However, Insel felt that the traditional diagnostic labels of psychiatry (like depression and schizophrenia) were more of an impediment than a help in understanding how the brain leads to mental illness. We already saw that research from both genetics and neuroscience has shown that those different diagnoses don't map well onto biology—just as one would not expect a symptom like chest pain to have a simple biological basis. Instead, he proposed that we study the different systems in the brain that might give rise to these symptoms, just as we study biological systems in order to understand other diseases. During his tenure as director (which ended when he moved to Google in 2015), Insel pushed the entire NIMH research program away from studying people based on their DSM diagnoses and more toward studying the basic brain systems that underlie those diagnoses. Groups of neuroscientists were convened to discuss each of a number of different domains that are thought to play a role in mental illness, with each group developing a "matrix" describing how that specific process is related to different genes, molecules, brain systems, and behaviors.

Fear in the Brain

One of the RDoC domains is "negative valence systems," which encompasses a set of negative emotional experiences including fear, anxiety, loss, and "frustrative nonreward." Of these, fear is probably the best understood in terms of the underlying brain systems. Fear is a natural emotion, acquired through evolution in order to help protect us from predators and prevent us from standing too close to the edge of a cliff. Most people experience fear from time to time, and some sensation-seeking people even relish it, but for other people fear can become debilitating. For a person suffering from post-traumatic stress disorder (PTSD), the most subtle reminder of a traumatic experience can trigger intense fear, and for someone with panic disorder the fear can often seem to come from nowhere. In some cases the concern about these attacks can lead a person to become *agoraphobic*, a term that translates from Latin as "fear of the marketplace" but for us refers to a fear of crowds or public places, that leads the person to avoid interactions with the world. Fortunately, my fear of public speaking never led me to avoid it, but for many people social anxiety severely limits their ability to achieve their professional or personal goals.

Neuroscience research has provided a detailed picture of how fear works in the brain. Say that you are walking down the street and someone approaches you with a gun and asks for your wallet and phone. Humans are not born with a fear of guns; we learn this from experience, and this learning occurs through changes in the connections between neurons in a brain circuit that includes the amygdala as well as a number of other areas connected to it. This circuit sends messages to the rest of our body that ultimately cause the unpleasant physical sensations that we experience, but they also can allow us to regulate the anxiety response, such as when we realize that the gun is actually a toy and the perpetrator is a friend playing a Halloween prank. Much of our knowledge of the detailed circuitry of the fear response has come from studies of nonhuman animals like rats and mice, but neuroimaging has also shown us what fear looks like in the human brain. One way that this has been studied is through what is called "fear conditioning." In this kind of experiment, the volunteers are presented with cues (such as pictures or sounds); some of these are presented alone, while others are followed by a mild electric shock. Over time, the individuals start to fear being shocked whenever they see the cues that were paired with shock, and it's possible to examine brain activity when the subjects are experiencing that fear. What this research has found is that when a person experiences fear of an impending shock, there is activity throughout the brain's fear circuit, which shows that the brain function of humans and rats is not so different when it comes to basic emotional experiences like fear. It is worth pointing out that there are still many questions about how fear works in the brain, and it is increasingly thought that there are different brain circuits involved in the conscious experience of fear and the physiological responses and behavioral responses (such as freezing or avoiding) that are associated with fear.

Fearing a shock in the fear conditioning experiment is normal, but what about when fear goes awry? Researchers think that what happens in PTSD is that the brain's fear system generalizes the fear response to otherwise nonthreatening cues in the world, and is unable to extinguish this response even in the face of experience showing that no harm will come from those cues. Several meta-analyses have shown that people with PTSD and anxiety disorders exhibit an exaggerated response in the amygdala and insula to negative emotional stimuli. It also appears that the strength of the response in the brain's fear circuit is related both to the likelihood of developing PTSD following a traumatic event and to the likelihood of successful treatment. One particularly interesting study recruited individuals from an emergency room following a traumatic event, and then used fMRI within about two months after the event to assess their brains' response to viewing fearful faces.[9] When they followed up with these individuals a year after the traumatic event, they found that people with higher activity in the amygdala within two months of the traumatic event were more likely to show symptoms of PTSD a year later. It is this kind of approach, focusing on the prediction of specific psychological symptoms using neuroscience, that RDoC has championed and that many hope will move psychiatry beyond its current messy state. At the same time, the past decade of research using the RDoC approach has not provided the level of breakthroughs that was initially hoped for, highlighting just how far we are from a deep understanding of these disorders.

Computational Psychiatry

The RDoC approach focuses on understanding which brain systems go awry in mental illness, but doesn't really tell us what specifically those brain systems are doing wrong in these disorders. To solve this problem, an emerging approach known as "computational psychiatry" is trying to link the brain dysfunction in mental illness back to the basic computations being performed by the brain, using sophisticated mathematical models.

Michael J. Frank is a neuroscientist at Brown University who has used computational models to study a number of different neurological and psychiatric disorders, and his work has been at the vanguard of the computational psychiatry movement. His research has focused on how people learn from either good or bad experiences, using reinforcement learning models like those that I described briefly in chapter 7. You may remember from the earlier discussion that dopamine plays a critical role in reinforcement learning. The role of dopamine in reinforcement learning comes at least in part from the effects that it has in the basal ganglia, which receive a major input from dopamine neurons. Specifically, the effects of dopamine on learning appear to come from its effects on changes in the connection strength of neurons in the basal ganglia. However, as you might have expected, it turns out that the dopamine story is much more complicated.

One complication is that there are different sets of neurons in the basal ganglia that respond differently to dopamine. One set of neurons, which Frank calls the "go pathway," causes us to engage in action when it is activated, while another set of neurons, called the "no-go pathway," drives us to avoid action when it is activated. Both of these sets of neurons are affected by dopamine, but in different and largely opposite ways, owing to the fact that they have different kinds of dopamine receptors. If neurons in the go pathway fire in the presence of dopamine, then the synaptic connections between the neurons and their inputs will be strengthened; conversely, if they fire in the absence of dopamine, those connections will be weakened. Exactly the opposite thing happens in the no-go pathway: firing in the presence of dopamine causes their connections to become weaker, and in the absence of dopamine to become stronger. Based on this, Frank had the idea that dopamine should have different effects on learning from positive outcomes (which should cause dopamine increases) and negative outcomes (which should cause dopamine decreases). His early work tested this idea in people with Parkinson's disease, who have dysfunctional dopamine systems. He found that unmedicated patients (who have low dopamine levels) had problems learning from positive outcomes, but not from negative outcomes; on the other hand, when they were medicated (which causes high dopamine levels), they showed the opposite pattern, having trouble learning from negative outcomes but not from positive outcomes.[10]

The ideas developed in Frank's early studies of Parkinson's disease have since been used to begin understanding how brain computations malfunction in people with schizophrenia.[11] Dopamine clearly plays an important role in schizophrenia— most of the drugs used to treat psychosis have their effects by blocking dopamine receptors—but an understanding of exactly what role it plays has eluded researchers. Frank and his colleagues have shown that individuals with schizophrenia behave similarly to unmedicated people with Parkinson's disease on their test: they are impaired at learning from positive feedback but normal at learning from negative feedback. Further, this impairment is related to the strength of so-called "negative symptoms" of schizophrenia, such as flattened affect (reduced expression of emotions), apathy, and a general lack of engagement with the world. The models that Frank and his colleagues have developed also suggest that part of the problem in schizophrenia might be related to *when* the dopamine neurons fire, proposing that dopamine neurons may fire more often at inappropriate times. This kind of computational analysis of behavior using brain-inspired mathematical models is likely to become a central feature of psychiatric research in the coming years, as the computational psychiatry movement gains steam. It remains to be seen whether it will help us understand psychiatric disorders better than our previous approaches have, but it's hard to imagine that it could do much worse.

Is Addiction a Brain Disease?

Addiction is a problem that touches almost everyone's life today in some way. Since my childhood, I have heard stories of one of my great-grandfathers whose alcoholism shattered the family. As an adult, I have had friends whose lives have been turned upside-down by their addiction to drugs, and have also watched as some of my favorite musicians and artists succumbed to their addictions. I'm not unique in this respect: across the world drug addiction takes a substantial toll. The most prevalent addiction is tobacco; it is estimated that more than 10% of deaths in men and 5% of deaths in women across the world are directly related to smoking. Addiction seems on its face to be different from other mental disorders because it is ultimately a disorder of choice: the addict could in principle simply choose to stop taking the drug, whereas it's much harder to imagine that psychotic people could choose not to listen to the voices in their head or depressed people could just will themselves into a good mood. But it's clear that stopping once one is addicted is very difficult; for example, studies of people trying to quit smoking or drinking have found that only about one-third of those who try to quit will still be successful a year later, and the outcomes for people addicted to opiates like heroin are usually much worse.

Addiction has been defined by the US National Institute on Drug Abuse (NIDA) as "a chronic, relapsing brain disease that is characterized by compulsive drug seeking and use, despite harmful consequences."[12] What does it mean for addiction to be a "brain disease"? There is a trivial sense in which addiction is a brain disease, since addicts' health is clearly impaired owing to their behavior, and all behavior arises from our brains, but the real story is much more complex. Rather than thinking of it as a disease of the same kind as schizophrenia, I tend to think of addiction as the result of a mismatch between our evolved brain and our modern environment.

In chapter 7 we discussed the brain's system for making choices, and the central role of dopamine in creating habits. All forms of addiction seem to involve dopamine; in fact, many of the most powerful drugs of addiction are those that directly affect the dopamine system in the brain. Cocaine causes a flood of dopamine in the synapses between neurons by turning off a chemical pump that usually sucks extra dopamine from the synapse back into the cell so that it can be recycled and used again. Amphetamines can actually cause these pumps to go in reverse, spewing out even more dopamine into the synapse. Other drugs of abuse (including alcohol and nicotine) have more indirect effects on dopamine, but ultimately it seems that dopamine is the key to all forms of addictive behavior. The fact that dopamine agonist medication for Parkinson's disease can cause strange addictions (as I mentioned in chapter 7) provides even stronger evidence for this idea. All of this is due to two of dopamine's main effects: it causes us to be motivated to obtain rewards ("wanting") and it increases the likelihood that any action that results in dopamine release will be repeated in the future, turning behaviors into habits.

Our brains evolved in a world where the strongest stimulation that our dopamine system ever received was probably from sexual intercourse. The foods that our hunter/gatherer ancestors ate were almost certainly healthier than those that most humans eat today, but it's doubtful that they were particularly tasty given that there was little access to salt, sugar, or spices, and whatever fruits they scavenged had not been bred for flavor like the supersweet strawberries and apples that we eat today. Fast-forward to modern society, and we now have access to an enormous number of ways to stimulate our dopamine system that go far beyond what evolution had prepared us for, from highly palatable junk foods to both legal and illegal drugs that stimulate dopamine release in ways that were unprecedented in our evolutionary history.

Neuroimaging research has shown that the brains of addicts exhibit specific changes in the dopamine system. Dopamine levels can't be measured using MRI, but they can be measured indirectly with PET imaging using radioactive tracers that are attached to a molecule that binds to dopamine receptors, such as raclopride. Several studies have found that there is decreased binding of these molecules in the brains of people addicted to stimulants such as cocaine. Because these molecules bind less strongly than real dopamine, they will only attach to receptors if there are no extra dopamine molecules available to take those spots. A lower level of binding thus could mean two things: more dopamine is present (taking up all the open spots), or there are fewer available receptors to latch on to. However, it's also possible to test how much dopamine is released in the brain by using the same PET imaging method while administering a drug that causes dopamine release; because of the difficulty of doing research with an illegal drug like cocaine, this has generally been done using other drugs with similar effects on dopamine, such as the prescription drug Ritalin (methylphenidate). Studies using this approach have found that drug abusers appear to show decreased release of dopamine when given these drugs, which suggests that their lower levels of binding probably indicate lower numbers of receptors present. This probably reflects the fact that, just like our bodies, our brains adapt to our circumstances, always trying to keep us within a range of healthy function. When the brain experiences abnormally high levels of dopamine, it adapts by both decreasing its response to dopamine (turning down the number of dopamine receptors) and reducing the amount of dopamine that it releases. This adaptation is probably part of why over time the abuser goes from feeling high from the drug to needing the drug just to feel normal.

The fact that drug use causes changes in the brain fits with the idea that addiction is a "disease," but it's worth noting that changes in the body due to experience happen for many reasons, not all of which we would think of as diseases. If I lift heavy weights regularly my muscles will grow, and if I eat too much junk food I will gain body fat, both of which are caused by the body's natural mechanisms for adapting to its environment. We certainly wouldn't refer to large biceps as a disease, and we only treat that extra fat as a disease if it gets too far out of hand.

Many drug treatment programs claim that drug addiction is a chronic disease that can be managed (usually through complete abstinence) but never cured, but there is some reason to question this idea. The most compelling evidence comes from a landmark study of heroin addicts returning from military service in Vietnam.[13] Narcotics were remarkably easy to obtain in Vietnam; almost half of the army's enlisted men in Vietnam in 1970–71 had tried narcotics, and 20% claimed to be addicted. If addiction is a chronic disease, then we would expect their addiction to continue when they returned to the United States after their service, but the data showed otherwise. A study by Lee Robins and her colleagues found that only about 5% of those soldiers who had been addicted in Vietnam had become readdicted a year after being back home. This was in stark contrast to the findings on drug addicts in the United States who had been sent to a "narcotics hospital," two-thirds of whom relapsed within six months. This tells us that context is very important, which is not surprising at all to anyone who has ever smoked; even if one can resist the impulse to smoke during the day, being in a smoky bar at night makes it much harder to say no. On the other hand, many of these individuals went on to develop other addictions (particularly alcohol), which suggests that they may have traded one addiction for another.

In fact, context is central not just to the desire to consume drugs but also to our brain's reaction to those drugs. When a person consumes a drug over time, the drug has increasingly weaker effects, which scientists call *tolerance* (and Australians call "piss fit"). The work of Shepard Siegel and his colleagues has shown that this tolerance occurs through learning and is specific to the context in which the drug is taken.[14] This fact explains an interesting aspect of heroin "overdoses"—in many cases, the amount of drug that results in death is no larger than the amount that the user has consumed in the past, but the drug is consumed in a different context than the user had previously consumed in. Siegel reported one particularly striking case of a cancer patient who received morphine at home for his pain. For about a month he had received his treatments in the bedroom, but one day decided to take the treatment in his living room instead. The dose was no different from what he had received previously, but in this new context he developed the signs of an overdose and died shortly thereafter. The context sensitivity of addiction helps to explain the low rate of readdiction in the returning Vietnam veterans, but also leads us to question the description of addiction as a "chronic disease."

The Stigma of Mental Illness

One motivation for treating addiction and other psychiatric illnesses as "brain diseases" is that it is thought that this might help alleviate the stigma associated with these disorders. Research into stigma has found that it can take three forms: fear that the mentally ill person will become violent and dangerous, concern that the individual is not responsible enough to make their own life choices, and feelings that the individual needs to be taken care of.[15] These societal stigmas can boomerang, leading mentally ill individuals to engage in "self-stigma" that can cause them to feel even worse about their disease. They can also lead to structural discrimination against mentally ill people; for example, in about one-third of states in the United States, people with mental illnesses may be denied the right to vote.

It's less common for people with physical diseases to be stigmatized (though it does sometimes happen, as in the case of HIV/AIDS), and one might hope that treating mental illness as a biological illness would decrease stigma. Studies of public perception of mental illness have confirmed that people across the world have become increasingly likely to consider major mental illnesses such as schizophrenia and depression as "brain diseases."[16] However, these changes in attitudes about the causes of mental illness have not necessarily translated into greater acceptance of mentally ill individuals; the same study that found increasing belief in biological causes for these illnesses also found that the willingness to accept a mentally ill person as a neighbor or coworker actually went *down* over the same period. According to the researchers Nick Haslam and Erlend Kvaale, the belief in what they call "biogenetic" explanations for mental illness has been a "mixed blessing": while it leads people to be less likely to blame the mentally ill for their problems, it actually makes them *more* likely to think that the mentally ill are dangerous, and also makes them more pessimistic about the ability of mentally ill people to improve through treatment.[17]

It's also interesting to look at how people with mental illnesses think about this question. Carla Meurk and her colleagues have done extensive interviews with people suffering from alcohol and drug addiction, delving specifically into the question of how they think of the idea of addiction as a brain disease. Their views are remarkably diverse. Some are clear in their rejection of the label:

> "I just don't think it could be a brain disease because it's something that it's by choice." [ID04, female]

Others embraced the label:

> "Well I'm happy if it is a disease because it helps me. Takes that pressure off—I'm a fuck up. [...] Ah I can say it's not just me and it's a disease. It's not just me being a screw up that is by drinking all the time, so yeah that made me feel better. [...] [If it's a disease] then at least I know all right there's a problem there and it can be fixed hopefully." [ID06, male]

Interestingly, some also raised concerns consistent with the mixed-blessings model that the disease label might lead them to be viewed as inferior:

> "Brain disease might sort of infer that—like brain injury. If you hear that somebody's got a brain injury or an acquired brain injury, you sort of get that image in your head of slow, low intelligence, as a result." [ID07, male][18]

Together, these studies show that the impacts of biogenetic explanations for mental illness and addiction are complex, and almost certainly there will be both positive and negative impacts on people suffering from those illnesses. Hopefully, with improved treatment for these disorders, the stigma will be reduced as they come to be viewed like other curable diseases.

CHAPTER 9

THE FUTURE OF NEUROIMAGING

We are clearly just in the early infancy of neuroimaging, and there is no doubt that our ability to decode the mind will become radically more powerful in the coming years. But how far can it go? In chapter 4 I laid out the idea of fMRI decoding in terms of translation between the language of humans and the language of the brain. This translation problem is particularly challenging because of the huge number of neurons in the brain—we need to listen to the activity of more than 10 billion neurons in order to make our translation. What we do in fMRI is to combine many neurons into each voxel, as if we have an intermediary translator who is summarizing into a single voice everything being said by the million or so neurons within the voxel. In addition, remember that fMRI is not listening directly to the neurons, but rather to the blood vessels that change in relation to the activity of neurons. Thus, our intermediate translators (voxels) are relaying a summary of the message after it has already been translated from the language of neurons to the language of blood flow. And those blood vessels are "slow talkers"—it's as if they listen to the neurons talk for a little while and then summarize what they have been talking about during that whole time. Finally, there are all sorts of artifacts that can get in the way of the translation, such as head motion or the effects of breathing, as if the translator gets distracted or burps occasionally. If this is starting to sound like a game of multilingual telephone with poor equipment, that's about right—not exactly a recipe for precise decoding of thoughts.

In chapter 4 I also mentioned the fallback idea of building a dictionary to map patterns of brain activity onto specific thoughts, but even here there are major challenges. As you have seen so far, we can decode what kind of object a person is seeing or thinking about with very high accuracy, at least at a broad stroke (such as cars versus houses, or addition versus subtraction). However, getting the finer details right remains out of our reach for the most part; we can decode with perfect accuracy whether you are looking at a face or an outdoor scene, but knowing *whose* face you are looking at remains surprisingly difficult. This probably reflects the fact that the distinction between faces and places relies upon voices coming from very different voxels, whereas the distinction between two different faces probably relies on finer differences between what neurons are saying within a specific voxel, and we just can't hear that with fMRI. The neuroscientist Luiz Pessoa of the University of Maryland also points out that any such "dictionary" of the brain will likely be highly contextualized; the specific meaning of any particular "word" in the dictionary will depend on the specific context in which it's being used—just as the meaning of the term "bank" depends on whether it's being used in the context of money, a river, or a an airplane, and the word "kill" can have almost contradictory meanings for a soldier and a stage performer.

We are now several years beyond the date when Marcel Just predicted that we should have full-blown fMRI mind reading in his interview with Lesley Stahl, yet we are still miles away from that goal. Neuroimaging using BOLD fMRI has told us a lot about how the human brain works, but it is also clearly limited: because the blood flow response that occurs in relation to neuronal activity is slow, we can't image brain activity at its actual time scale, but rather must settle for a blurry average over much longer periods of time. The blood flow response is also spatially blurry, because standard fMRI techniques are primarily sensitive to oxygenation changes in the blood vessels that flow out of the cortex, which can happen up to several millimeters from the actual neuronal activity. Thus, even if we could create fMRI images with much smaller voxels, it wouldn't help, because the blood flow response is going to be a blurry lens through which to view brain activity—it would be like buying a digital camera with more megapixels but then taking photos through a foggy lens. These limitations have led researchers to try to find methods that might give us a clearer view of brain activity.

This One Goes to 11—MRI at Higher Magnetic Fields

One hope is that the use of MRI scanners with stronger magnetic fields could provide fMRI images with better resolution. The standard hospital MRI scanner has a magnet with a strength of 1.5 tesla, and research scanners in most centers have a strength of 3 tesla, but increasingly MRI centers are installing scanners with much higher field strength. There are more than 60 scanners with a strength of 7 tesla installed across the world, and an operational 9.4-tesla scanner in Maastricht in the Netherlands— which was the strongest scanner approved for human imaging, until a 10.5-tesla scanner at The University of Minnesota (see figure 9.1) was approved for human scanning in 2018. An 11.7-tesla human scanner was installed at the NIH in 2012, but before it could be used to scan humans it experienced a damaging *quench*; this occurs when the superconducting coil of the magnet loses its superconductivity, which results in a rapid increase in temperature that causes the liquid helium coolant to boil. The increase in pressure associated with this boiling causes the gas to be explosively released from the scanner through a specially designed safety vent. Quenching is very rare and often results in damage to the magnet; in this case the magnet was ruined, and it has yet to be replaced.

Imaging at higher fields is useful not just because it allows the creation of images with higher spatial resolution (often less than one millimeter), but also because it allows the use of different fMRI techniques that are more sensitive to blood oxygenation changes happening in the smaller blood vessels that are much closer to where the neuronal activity is actually happening. One of the most compelling demonstrations of the usefulness of high-field MRI for precise mapping of the brain was presented by Essa Yacoub and his colleagues from the University of Minnesota, who examined the ability to image a feature of the visual cortex

Figure 9.1. An image of the 110-ton 10.5-tesla magnet as it was being installed at the Center for Magnetic Resonance Research (CMRR) at the University of Minnesota. Standing in front of the scanner is Kâmil Uğurbil, director of the CMRR and one of the early pioneers of fMRI. Image courtesy of Kâmil Uğurbil.

known as *orientation columns*.[1] These are small columns of cells in the visual cortex less than a millimeter wide that respond preferentially to lines of a particular spatial orientation; across the visual cortex they are arranged in a stereotyped pattern that has been characterized in detail in studies of nonhuman animals. Studies using three-tesla MRI have not been able to localize these orientation columns in humans, presumably because the blood flow patterns are too broad to precisely image the small columns. However, using a technique known as *spin echo* MRI, Yacoub and colleagues were able to map these orientation columns and relate their characteristics to the known features from animal studies. Subsequent work by Jon Polimeni and his colleagues at the Massachusetts General Hospital has also shown how seven-tesla MRI can be used to map the response in different layers of the cerebral cortex. Because the different layers play different roles in the brain's computations (for example, with some layers receiving input from other brain areas versus sending output), the ability to image these *laminar* patterns of activity represents a major advance in the development of fMRI techniques. This research shows how the higher resolution obtained using high-field imaging opens up a whole new set of questions, by allowing researchers to examine brain organization at the level of cortical columns and layers rather than simply looking across different regions of the brain.

Higher-field MRI scanners are likely to become increasingly available since the approval in 2017 of 7-tesla MRI by the US Food and Drug Administration for routine clinical use. However, despite its promise, high-field fMRI remains challenging. One practical limitation is the need for huge amounts of helium to cool the monstrous magnets needed for high-field human imaging; owing to the ongoing worldwide helium shortage, it took more than a year for the University of Minnesota to obtain enough helium to cool their 10.5-tesla scanner when it was installed in 2014. Another limitation relates to safety. The radio frequency pulses that are used to generate MRI images can result in the heating of tissue within the subject, and higher magnetic fields require a greater amount of radio frequency energy to reach the middle of the brain. MRI scanners have built-in safety mechanisms that prevent heating, and these mechanisms will likely limit scanning in ways that could prevent high-field MRI from reaching its full theoretical potential. A more fundamental limitation is that high-field fMRI mostly still relies on imaging the blood flow or oxygen changes, which are slow. Thus, while it may be able to improve the spatial resolution of fMRI, it will not be able to image brain responses on the same time scale that they occur in neurons. Finally, imaging at higher magnetic fields leads to increased artifacts in the images, making it difficult to image some parts of the brain.

Beyond BOLD

Given the fundamental limitations of imaging the BOLD response, yet another hope is that there may be other ways of using MRI to more directly measure neuronal activity. One method that has been proposed is to measure the changes in the movement of water molecules (known as *diffusion*) that occur when neurons fire, owing to the fact that neurons swell very slightly when they fire. There is some debate over whether diffusion-based fMRI actually works, and even if it does the signals are very small—much smaller than the changes that we can see using BOLD fMRI, meaning that it would be very difficult to measure subtle effects without very large samples. In addition, because diffusion MRI is measuring microscopic movements of water molecules, it is highly sensitive to movements of the head. There has also been enthusiasm in the past about being able to measure electrical activity in neurons with MRI, called *neuronal current imaging*. This approach attempts to measure the effects on the MRI signal of the minute electrical currents that occur in neurons as they fire; while it has potential to greatly improve the ability to image brain function, the signals that can be obtained from these measurements are drastically smaller than those obtained using standard BOLD fMRI, so they have not yet been applied to neuroscience questions.

Another alternative would be to measure electrical activity directly. There are well-established methods for this, which I have not said much about in this book. These include EEG, which measures electrical activity using electrodes placed on the scalp, and magnetoencephalography (MEG), which measures the minute magnetic fields that emerge from electrical activity in the brain. Those techniques can, in principle, provide us with a measurement of brain activity that is more direct than the measure of blood oxygenation that we get from BOLD fMRI. However, in practice they have limits that have prevented them from having the same impact that fMRI has had on the field of cognitive neuroscience. The reason is simple: while EEG and MEG can provide us with measurements that have very high temporal resolution (at the level of milliseconds), our ability to see *where* the activity is coming from is extremely limited. This is especially true for EEG, where the electrical activity of the brain has to pass through the skull and scalp before we can measure it, which blurs the signals so badly that it is impossible to tell for sure where they are coming from. Researchers in that field have developed ways to try to solve this problem, but most of them rely on assumptions about the signals that many researchers find simply too strong, leaving the results open to question. MEG fares a bit better on this account, since its signals are not blurred in the same way, but it suffers from the problem that it can only see activity in parts of the brain that are arranged in a certain way—because of how magnetic fields are generated, MEG can only see signals coming from bits of the cerebral cortex that are aligned at a right angle to the surface of the skull. There has long been an interest in combining EEG or MEG data with fMRI data, but so far this approach has not been very successful, in part because it is highly challenging from a technical standpoint. Despite these technical challenges, the combination of fMRI with EEG and/or MEG is currently the most promising next step in neuroimaging.

One technique that has gained some prominence for brain imaging uses near-infrared light to measure the same changes in blood oxygenation that are measured by BOLD fMRI. Known as near-infrared spectroscopy (NIRS), this technique places a set of lasers against the scalp, alongside a set of light sensors that measure the light that comes back out. Oxygenated blood reflects light differently from deoxygenated blood—the reason that arterial blood appears red and venous blood appears blue—and these differences can be measured in the light that emerges back from the brain. The main benefit of NIRS is that it is much less expensive than fMRI, and also very portable, allowing it to be used in a variety of real-world situations where fMRI would not be possible. However, in some ways it inherits the worst of both fMRI and EEG/MEG: it has poor temporal resolution because it is measuring the same blood flow response that is measured by fMRI, and it has poor spatial resolution because the skull blurs the light coming back from the brain. Perhaps the most promising application of NIRS is in the imaging of very young infants, whose skulls are very thin and who also would be very difficult to image using fMRI. However, I personally don't see NIRS playing a major role in cognitive neuroscience in the future.

There may yet be other ways of imaging human brain activity that can overcome the limitations of these existing techniques, but this imaging is fundamentally limited by the fact that we generally must look at the brain from the outside, except for the rare cases of surgical patients like the one I described in chapter 1. By contrast, the ability of neuroscientists to image brain activity in nonhuman animals has grown immensely in the past decade, largely owing to the use of molecular biology and genetics techniques to transform various aspects of neuronal activity into light. At present, it is possible to image the activity of many, if not all, of the neurons in simple organisms such as worms, flies, and the tiny zebra fish, all of which are favorites of neuroscientists because of the ease with which they can measure and manipulate their brain activity. Within the next decade it will probably be possible to measure the activity of nearly all of the neurons in the brain of a mouse or rat as well. Unfortunately it is unlikely that these techniques will ever be suitable for humans, because they generally require the animals to be genetically engineered, and they also require that the skull be removed so that the brain can be observed directly. Despite this, we can still learn a lot from studies of animals, and much of the research that I have outlined in this book has benefited from research that has more directly studied brain activity in nonhuman animals. It's only by this combination of techniques across species that we will likely ever gain a handle on how the brain works.

Improving Science through Transparency

Throughout this book I have pointed to problems with the way that scientific research has been done in the past, which has in some cases led to research results that cannot be reproduced. These problems are not unique to neuroimaging, but some of the features of neuroimaging research, such as the large amount of flexibility in how the data are analyzed, make it particularly susceptible to those problems. The ability to reproduce others' results is the defining characteristic of science, and the demonstration that research practices have led to irreproducible results has driven some scientists to intensely reexamine our approaches. A number of different ideas have emerged about how to improve science, some of which I have already discussed, such as the idea of "preregistration" in which the researcher provides a detailed description of the analyses and hypotheses that will be tested prior to ever touching the data. These changes are certainly important, but I and others have argued that in order for science to improve we also need to adopt a kind of radical transparency, particularly through the open sharing of data.

In some areas of science the sharing of data is commonplace, such as data collected from unique, expensive instruments like the Hubble Space Telescope. However, in many parts of the life sciences, the sharing of data is much less common. Most scientific journals require that researchers commit to sharing their data upon request, but there are generally no teeth to these policies, and it is common for researchers to drag their feet when asked to share their data, or simply not respond to requests at all. There are many reasons for this, but probably the most common is that researchers want to retain an advantage over their competitors; if they distribute the data, then other researchers may be able to test hypotheses that the data collectors were planning to test themselves, and might end up beating the data collectors into print with their results, which we refer to as "being scooped." This matters to researchers because there are strong incentives to publish their papers in high-profile journals, and those journals place a high premium on "novelty"—that is, they want to publish the first paper that demonstrates some particular finding. A paper that simply replicates findings published previously, while very important for science, will often end up being published in a lower-profile journal, and this can have career implications for the researcher, since jobs and promotions usually go to the researchers who are publishing in high-profile journals. The debate over the sharing of data became particularly heated in 2016, when an editorial published in the *New England Journal of Medicine* used the term "research parasites" to describe scientists who use data collected by other researchers rather than collecting their own data.[2] The editorial resulted in a firestorm of criticism, and the authors ultimately published another editorial that backtracked on their criticism, but it nonetheless laid bare the true feelings of some researchers about data sharing.

The field of neuroimaging has actually been at the vanguard of data sharing, though not without an initial false start. In 1999, Michael Gazzaniga started a project called the fMRI Data Center at Dartmouth University, which was meant to share data from published fMRI studies. Gazzaniga was the editor of the *Journal of Cognitive Neuroscience*, one of the most prominent journals for cognitive neuroscience research, and he used this position to mandate that all researchers publishing fMRI studies in the journal would be required to share their data via the fMRI Data Center. Unfortunately the project was too far ahead of its time, and many researchers had not yet bought into the idea of data sharing. An open letter was circulated by a group of fMRI researchers that strongly protested the requirement for data sharing, and while the fMRI Data Center would ultimately share data from more than 100 published studies, it remained somewhat marginal until it finally closed in 2012 owing to lack of funding.

It was another project that reignited the ongoing data sharing revolution in neuroimaging. In 2009, an international group of researchers released resting fMRI data collected on more than 1,300 individuals at 30 different sites around the world, in what was known as the 1,000 Functional Connectomes Project.[3] These data were used in a number of research studies, and the success of this project has shown the way for several other large data sharing efforts. Perhaps the most successful has been the Human Connectome Project (mentioned in chapter 3), which was led by David Van Essen of Washington University in St. Louis and Kâmil Uğurbil from the University of Minnesota. This project received $30 million from the US NIH with the primary goal of collecting and sharing a highly advanced data set on 1,200 individuals that would characterize the connectivity of the human brain. As I mentioned before, these data have been remarkably useful, contributing to hundreds of research articles to date. An even larger study is ongoing in the United Kingdom, known as the UK Biobank. This project plans to scan the brains (as well as other organs) of 100,000 UK citizens, and all of the data will be made available to researchers around the world.[4] My part in the data sharing enterprise was to start a project called OpenfMRI in 2010 that aimed to carry on the legacy of the fMRI Data Center.[5] This project focused primarily on sharing data from fMRI studies that used psychological tasks to image brain activity, as opposed to the 1,000 Functional Connectomes Project that shared data collected from people at rest. OpenfMRI has also been very successful, with well over 100 publications that have reused the data; we have estimated that this data reuse has saved researchers more than $3 million, and it has provided researchers around the world with data that they can use to ask new questions, even if they don't have access to an MRI scanner, in addition to allowing researchers to check one another's work.

One decision that I made at the outset of the MyConnectome study was that nearly all of the data would be released publicly, including neuroimaging, genetics, and psychological tests.[6] Usually when we share data they are "de-identified," meaning that we remove any information that might allow someone to figure out who the data came from—this includes removing features from images that might let someone reconstruct the person's face. This helps protect the privacy of research subjects, which is a promise that we make to them when they participate. In my case, however, it would be impossible to de-identify the data, since everyone in the field would know who the data came from. Opening one's self up in this way certainly comes with risks. The risks of sharing my genetics data are relatively low, because of the legal protections afforded by the Genetic Information Nondiscrimination Act, which protects individuals in the United States from discrimination based on genetic information. However, my other information comes with no such protection, so it is possible that I might be denied insurance one day because the company doesn't like something that it sees in my brain or behavior. I took the risk because I think that progress in medicine will require everyone to become more open to sharing their data. As a tenured professor at an elite university, I am also in a very privileged position, so I figured that if anyone could afford to take the chance, I could.

The data from my study have already had a remarkable impact, with more than 10 scientific papers so far that have used the data since they were released in 2015. Several of these papers have used the data to test out new methods for MRI data analysis, because they provide a uniquely powerful resource for testing such methods. Others have used them to establish new relations between brain function and mental state. Research by Rick Betzel and Dani Bassett from the University of Pennsylvania analyzed how brain function changed in relation to my mood across different days, and found that a specific aspect of brain connectivity known as "network flexibility" was greater on days when I was in a better mood.[7] It's now common to see my brain in presentations at scientific meetings, and at a recent meeting someone I didn't know approached me at the poster session and said "Hi Dr. Poldrack, I'd like to show you our poster, it has your brain in it." My sharing has also inspired others to share their data; most recently, a group of researchers from Washington University led by Dr. Nico Dosenbach have released a data set that they call the "Midnight Scan Club," so named because it was collected from a set of researchers who got into the MRI scanner repeatedly late at night. This study has already let them confirm some of the findings from the MyConnectome study, showing that each individual has idiosyncratic features of brain organization similar to those that we found in my brain.

Scientists can often be their own worst enemies; by continually criticizing ourselves, we run the risk of delegitimizing our enterprise in the eyes of the public, who fund our research and rely on our results to make many important decisions. It is important to remember that science is not simply a body of knowledge, but rather a way of asking questions. It's not guaranteed to give us the right answer, and in fact we are pretty sure that many of the ideas we hold right now about how the brain works will be overturned in the long run, but the importance of the scientific method is what it drives us to do: rather than asking how we can find support for our ideas, science drives us to find out how we might be wrong, and the constant questioning and self-doubt are ultimately the best way that we know to keep from fooling ourselves.

Conclusion

When I first learned about neuroimaging in the early 1990s, I was skeptical to say the least. How could the measurement of oxygen levels in the blood tell us about the intricate operations of the brain and their relation to mental function? In the ensuing 25 years, those early concerns were largely allayed, but new concerns have emerged, as you have seen throughout this book. In addition to questions of basic science, a whole new set of questions has arisen around applications of neuroimaging to real-world questions. Ultimately fMRI has been incredibly useful and has told us much about the function of the human brain. We now understand much more about the way that mental function is organized across the brain, and how different brain regions work together to provide us with many of the cognitive abilities that make us human. We have also begun to understand what makes each individual human unique, and how the human brain changes over many different time scales, from seconds to years. Neuroimaging has provided a new window into the brain dysfunctions that lead to mental illness, and in fact is driving a wholesale rethinking of how we conceptualize mental illness. And the ability of fMRI to decode mental states leads us to ask deep questions about the relation between the mind and the brain and what this means for our individual identities as humans.

Despite the progress made with fMRI, every scientific technique has its day, and ultimately there likely will be new technologies that will allow us to measure brain function with much higher temporal and spatial resolution than we can with current fMRI techniques. Whether that will happen within my lifetime is not a prediction that I would want to put money on, but perhaps my ventral striatum knows better than I do.

NOTES

Chapter 1

1. D. F. Rolfe and G. C. Brown, "Cellular Energy Utilization and Molecular Origin of Standard Metabolic Rate in Mammals," *Physiological Reviews* 77, no. 3 (1997): 731–58.
2. R. Douglas Fields, Alfonso Araque, Heidi Johansen-Berg, Soo-Siang Lim, Gary Lynch, Klaus-Armin Nave, Maiken Nedergaard, Ray Perez, Terrence Sejnowski, and Hiroaki Wake, "Glial Biology in Learning and Cognition," *Neuroscientist* 20, no. 5 (2014): 426–31, doi:10.1177/1073858413504465.
3. The names of some individuals have been changed to maintain their privacy.
4. Josef Parvizi, Corentin Jacques, Brett L. Foster, Nathan Withoft, Vinitha Rangarajan, Kevin S. Weiner, and Kalanit Grill-Spector, "Electrical Stimulation of Human Fusiform Face-Selective Regions Distorts Face Perception," *Journal of Neuroscience* 32, no. 43 (2012): 14915–20, doi:10.1523/JNEUROSCI.2609-12.2012. Quote from Movie 1, time point 0:10.
5. Henry L. Roediger and Jeffrey D. Karpicke, "Test-Enhanced Learning: Taking Memory Tests Improves Long-Term Retention," *Psychological Science* 17, no. 3 (2006): 249–55, doi:10.1111/j.1467-9280.2006.01693.x.
6. Justin S. Feinstein, Ralph Adolphs, Antonio R. Damasio, and Daniel Tranel, "The Human Amygdala and the Induction and Experience of Fear," *Current Biology* 21, no. 1 (2011): 34–38, doi:10.1016/j.cub.2010.11.042.
7. Ibid., 35.
8. Marco Iacoboni, Joshua Freedman, Jonas Kaplan, Kathleen Hall Jamieson, Tom Freedman, Bill Knapp, and Kathryn Fitzgerald, "This Is Your Brain on Politics," *New York Times*, November 11, 2007, http://www.nytimes.com/2007/11/11/opinion/11freedman.html.
9. Russell A. Poldrack, "Can Cognitive Processes Be Inferred from Neuroimaging Data?" *Trends in Cognitive Sciences* 10, no. 2 (2006): 59–63, doi:10.1016/j.tics.2005.12.004.
10. Tal Yarkoni, Russell A Poldrack, Thomas E. Nichols, David C. Van Essen, and Tor D. Wager, "Large-Scale Automated Synthesis of Human Functional Neuroimaging Data," *Nature Methods* 8, no. 8 (2011): 665–70, doi:10.1038/nmeth.1635.

Chapter 2

1. David T. Field and Laura A. Inman, "Weighing Brain Activity with the Balance: A Contemporary Replication of Angelo Mosso's Historical Experiment," *Brain* 137, no. 2 (2014): 634–39, doi:10.1093/brain/awt352.
2. C. S. Roy and C. S. Sherrington, "On the Regulation of the Blood-Supply of the Brain," *Journal of Physiology* 11, no. 1–2 (1890): [85]–108, 158–17.
3. PET was preceded by another set of imaging methods developed by the Scandinavian

researchers David Ingvar and Niels Lassen in the 1960s. However, their methods were considerably less powerful than the PET imaging, and did not gain wide usage.

4. A fascinating autobiography outlining Raichle's career can be found at Marcus E. Raichle, Sr., *Marcus E. Raichle, Sr.* (Society for Neuroscience, n.d.), https://www.sfn.org /-/media/SfN/Documents/TheHistoryofNeuroscience/Volume-8/MarcusRaichle.ashx.

5. Michael I. Posner, *Michael I. Posner*, 564 (Society for Neuroscience, n.d.), https://www .sfn.org/-/media/SfN/Documents/TheHistoryofNeuroscience/Volume-7/c13.ashx.

6. John F. McDonnell, *Tribute to James S. McDonnell* (Washington, DC: National Academy of Sciences, 1999), https://www.jsmf.org/about/tribute.pdf.

7. Michael Posner, "Michael I. Posner," in *The History of Neuroscience in Autobiography*, vol. 7, edited by L. R. Squire, 576 (New York: Oxford University Press, 2012).

8. M. I. Posner, S. E. Petersen, P. T. Fox, and M. E. Raichle, "Localization of Cognitive Operations in the Human Brain," *Science* 240, no. 4859 (1988): 1627–31; S. E. Petersen, P. T. Fox, M. I. Posner, M. Mintun, M. E. Raichle, and P. T. Fox, "Positron Emission Tomographic Studies of the Cortical Anatomy of Single-Word Processing," *Nature* 331, no. 6157 (1988): 585–89, doi:10.1038/331585a0.

9. Gary Boas, "The Life and Science of Jack Belliveau: An Oral History," *FMRI25*, October 3, 2016, http://www.fmri25.org/life-science-jack-belliveau-oral-history/.

10. Benedict Carey, "Jack Belliveau, Explorer of the Brain Using M.R.I., Dies at 55," *New York Times*, March 9, 2014, https://www.nytimes.com/2014/03/10/science/jack -belliveau-explorer-of-the-brain-dies-at-55.html.

11. S. Ogawa, T. M. Lee, A. R. Kay, and D. W. Tank, "Brain Magnetic Resonance Imaging with Contrast Dependent on Blood Oxygenation," *Proceedings of the National Academy of Sciences of the USA* 87, no. 24 (1990): 9868–72.

12. Kenneth K. Kwong, "Record of a Single fMRI Experiment in May of 1991," *Neuroimage* 62, no. 2 (2012): 610–12, doi:10.1016/j.neuroimage.2011.07.089.

13. Peter A. Bandettini, "Sewer Pipe, Wire, Epoxy, and Finger Tapping: The Start of fMRI at the Medical College of Wisconsin," *Neuroimage* 62, no. 2 (2012): 622, doi:10.1016/j.neuroimage.2011.10.044.

14. The abbreviation PNAS is often jokingly translated as "post Nature and Science," reflecting the fact that it is often the journal where papers are submitted after having been rejected at *Nature* or *Science*, the most prestigious general science journals.

15. Kâmil Uğurbil, "Development of Functional Imaging in the Human Brain (fMRI): The University of Minnesota Experience," *Neuroimage* 62, no. 2 (2012): 613–19, doi: 10.1016/j.neuroimage.2012.01.135.

Chapter 3

1. N. K. Logothetis, J. Pauls, M. Augath, T. Trinath, and A. Oeltermann, "Neurophysiological Investigation of the Basis of the fMRI Signal," *Nature* 412, no. 6843 (2001): 150–57, doi: 10.1038/35084005.

2. Jin Hyung Lee, Remy Durand, Viviana Gradinaru, Feng Zhang, Inbal Goshen, Dae-Shik Kim, Lief E. Fenno, Charu Ramakrishnan, and Karl Deisseroth, "Global and Local fMRI Signals Driven by Neurons Defined Optogenetically by Type and Wiring," *Nature* 465, no. 7299 (2010): 788–92, doi: 10.1038/nature09108.

3. The field of neuroscience lost a rising star when Sergent took her own life in 1994, after being anonymously accused of research fraud, which was never substantiated.

4. Nancy has more recently become famous for shaving her head on live video in service of a brain anatomy lesson: Nancy Kanwisher, "The Neuroanatomy Lesson," *YouTube*, April 14, 2015, https://www.youtube.com/watch?v=PcbSQxJ7UrU.

5. N. Kanwisher, J. McDermott, and M. M. Chun, "The Fusiform Face Area: A Module in Human Extrastriate Cortex Specialized for Face Perception," *Journal of Neuroscience* 17, no. 11 (1997): 4302–11.

6. James V. Haxby, "Multivariate Pattern Analysis of fMRI: The Early Beginnings," *Neuroimage* 62, no. 2 (2012): 852–55, doi: 10.1016/j.neuroimage.2012.03.016.

7. J. V. Haxby, M. I. Gobbini, M. L. Furey, A. Ishai, J. L. Schouten, and P. Pietrini, "Distributed and Overlapping Representations of Faces and Objects in Ventral Temporal Cortex," *Science* 293, no. 5539 (2001): 2425–30, doi: 10.1126/science.1063736.

8. Haxby apparently didn't know this at the time, but there is an official name for this technique: the one-nearest-neighbor classifier.

9. Bharat B. Biswal, "Resting State fMRI: A Personal History," *Neuroimage* 62, no. 2 (2012): 938–44, doi: 10.1016/j.neuroimage.2012.01.090.

10. For a particularly approachable introduction this field, I recommend Duncan J. Watts, *Six Degrees: The Science of a Connected Age* (New York: Norton, 2003).

11. See for yourself at https://oracleofbacon.org/.

12. For a copy of their original poster, see Craig M. Bennett, Abigail A. Baird, Michael B. Miller, and George L. Wolford, *Neural Correlates of Interspecies Perspective Taking in the Post-Mortem Atlantic Salmon: An Argument for Multiple Comparisons Correction* (Prefrontal.org, 2009), http://prefrontal.org/files/posters/Bennett-Salmon-2009.pdf.

13. Sabrina M. Tom, Craig R. Fox, Christopher Trepel, and Russell A. Poldrack, "The Neural Basis of Loss Aversion in Decision-Making under Risk," *Science* 315, no. 5811 (2007): 515–18, doi: 10.1126/science.1134239.

14. Richard Feynman, *Cargo Cult Science*, commencement address (Caltech, 1974), http://calteches.library.caltech.edu/51/2/CargoCult.htm.

15. Russell A. Poldrack, Chris I. Baker, Joke Durnez, Krzysztof J. Gorgolewski, Paul M. Matthews, Marcus R. Munafò, Thomas E. Nichols, Jean-Baptiste Poline, Edward Vul, and Tal Yarkoni, "Scanning the Horizon: Towards Transparent and Reproducible Neuroimaging Research," *Nature Reviews Neuroscience* 18, no. 2 (2017): 115–26, doi: 10.1038/nrn.2016.167.

16. The title of the paper was later changed to the less incendiary "Puzzlingly High Correlations in fMRI Studies of Emotion, Personality, and Social Cognition." E. Vul, C. Harris, P. Winkielman, and H. Pashler, *Perspectives on Psychological Science* 4, no. 3 (2009): 274–90.

17. Russell A. Poldrack and Jeanette A. Mumford, "Independence in ROI Analysis: Where Is the Voodoo?" *Social Cognitive and Affective Neuroscience* 4, no. 2 (2009): 208–13, doi: 10.1093/scan/nsp011.

18. Matthew D. Lieberman, Elliot T. Berkman, and Tor D. Wager, "Correlations in Social Neuroscience Aren't Voodoo: Commentary on Vul et al. (2009)," *Perspectives on Psychological Science* 4, no. 3 (2009): 299, doi: 10.1111/j.1745-6924.2009.01128.x.

19. E. Vul, C. Harris, P. Winkielman, and H. Pashler, "Reply to Comments on 'Puzzlingly High Correlations in fMRI Studies of Emotion, Personality, and Social Cognition'," *Perspectives on Psychological Science* 4, no. 3 (2009): 319–24, doi: 10.1111/j.1745-6924.2009.01132.x.

Chapter 4

1. CBS, "Reading Your Mind," *YouTube*, January 4, 2009, https://www.youtube .com/watch?v=8jc8URRxPIg. Conversation starts at time point 12:44.

2. Bob McDonald, "Neuro-privacy," *Quirks & Quarks*, CBC Radio, May 13, 2017, 12.00 a.m. ET, http://www.cbc.ca/radio/quirks/neuro-privacy-1.4116070. Quote at time point 1:50.

3. John-Dylan Haynes and Geraint Rees, "Predicting the Stream of Consciousness from Activity in Human Visual Cortex," *Current Biology* 15, no. 14 (2005): 1301–7, doi: 10.1016/j.cub.2005.06.026.

4. Kendrick N. Kay, Thomas Naselaris, Ryan J. Prenger, and Jack L. Gallant, "Identifying Natural Images from Human Brain Activity," *Nature* 452, no. 7185 (2008): 352–55, doi: 10.1038/nature06713.

5. Yoichi Miyawaki, Hajime Uchida, Okito Yamashita, Masaaki Sato, Yusuke Morito, Hiroki C. Tanabe, Norihiro Sadato, and Yukiyasu Kamitani, "Visual Image Reconstruction from Human Brain Activity Using a Combination of Multiscale Local Image Decoders," *Neuron* 60, no. 15 (2008): 915–29, doi:10.1016/j.neuron.2008.11.004.

6. Thomas Naselaris, Ryan J. Prenger, Kendrick N. Kay, Michael Oliver, Jack L. Gallant, "Bayesian Reconstruction of Natural Images from Human Brain Activity," *Neuron* 63, no. 6 (2009): 902–15, doi:10.1016/j.neuron.2009.09.006.

7. Adrian M. Owen, Martin R. Coleman, Melanie Boly, Matthew H. Davis, Steven Laureys, and John D. Pickard, "Detecting Awareness in the Vegetative State," *Science* 313, no. 5792 (2006): 1402, doi:10.1126/science.1130197.

8. Martin M. Monti, Audrey Vanhaudenhuyse, Martin R. Coleman, Melanie Boly, John D. Pickard, Luaba Tshibanda, Adrian M. Owen, and Steven Laureys, "Willful Modulation of Brain Activity in Disorders of Consciousness," *New England Journal of Medicine* 362, no. 7 (2010): 579–89, doi:10.1056/NEJMoa0905370.

9. Marie-Aurélie Bruno, Jan L. Bernheim, Didier Ledoux, Frédéric Pellas, Athena Demertzi, and Steven Laureys, "A Survey on Self-Assessed Well-Being in a Cohort of Chronic Locked-In Syndrome Patients: Happy Majority, Miserable Minority," *BMJ Open* 1 (2011): e000039, doi:10.1136/bmjopen-2010-000039.

10. Mo Costandi, "Uncomfortably Numb: The People Who Feel No Pain," *Guardian*, May 25, 2015, https://www.theguardian.com/science/neurophilosophy/2015/may/25/the -people-who-feel-no-pain.

11. Tor D. Wager, Lauren Y. Atlas, Martin A. Lindquist, Mathieu Roy, Choong-Wan Woo, and Ethan Kross, "An fMRI-Based Neurologic Signature of Physical Pain," *New England Journal of Medicine* 368, no. 15 (2013): 1388–97, doi:10.1056/ NEJMoa1204471.

12. Javeria A. Hashmi, Marwan N. Baliki, Lejian Huang, Alex T. Baria, Souraya Torbey, Kristina M. Hermann, Thomas J. Schnitzer, and A. Vania Apkarian, "Shape Shifting Pain: Chronification of Back Pain Shifts Brain Representation from Nociceptive to Emotional Circuits," *Brain* 136, no. 9 (2013): 2751–68, doi:10.1093/brain/awt211.

13. Ashley Bell, "Neuroscience in the Courtroom: Can We Measure Pain?" *Law Street*, March 13, 2015, https://lawstreetmedia.com/issues/health-science/neuroscience- courtroom-can-measure-pain/.

14. D. Callan, L. Mills, C. Nott, R. England, and S. England, "A Tool for Classifying Individuals with Chronic Back Pain: Using Multivariate Pattern Analysis with

Functional Magnetic Resonance Imaging Data," *PLoS One* 9, no. 6 (2014): e98007, doi:10.1371/journal.pone.0098007.

15. Gaël Varoquaux, "Cross-Validation Failure: Small Sample Sizes Lead to Large Error Bars," *NeuroImage* (June 24, 2017), doi:10.1016/j.neuroimage.2017.06.061.

16. Choong-Wan Woo, Mathieu Roy, Jason T. Buhle, and Tor D. Wager, "Distinct Brain Systems Mediate the Effects of Nociceptive Input and Self-Regulation on Pain," *PLoS Biology* 13, no. 1 (2015): e1002036, doi:10.1371/journal.pbio.1002036.

Chapter 5

1. Bruce A. Yankner, Tao Lu, and Patrick Loerch, "The Aging Brain," *Annual Review of Pathology* 3 (2008): 41–66, doi:10.1146/annurev.pathmechdis.2.010506.092044.

2. Randy L. Buckner, Abraham Z. Snyder, Benjamin J. Shannon, Gina LaRossa, Rimmon Sachs, Anthony F. Fotenos, Yvette I. Sheline, William E. Klunk, Chester A. Mathis, John C. Morris, et al., "Molecular, Structural, and Functional Characterization of Alzheimer's Disease: Evidence for a Relationship between Default Activity, Amyloid, and Memory," *Journal of Neuroscience* 25, no. 34 (2005): 7709–17, doi:10.1523/JNEUROSCI.2177-05.2005.

3. A. M. Clare Kelly and Hugh Garavan, "Human Functional Neuroimaging of Brain Changes Associated with Practice," *Cerebral Cortex* 15, no. 8 (2005): 1089–102, doi:10.1093/cercor/bhi005.

4. Tal Yarkoni, Deanna M. Barch, Jeremy R. Gray, Thomas E. Conturo, and Todd S. Braver, "BOLD Correlates of Trial-by-Trial Reaction Time Variability in Gray and White Matter: A Multi-study fMRI Analysis," *PLoS One* 4, no. 1 (2009): e4257, doi:10.1371/journal.pone.0004257.

5. Jedediah M. Bopp, David J. Miklowitz, Guy M. Goodwin, Will Stevens, Jennifer M. Rendell, and John R. Geddes, "The Longitudinal Course of Bipolar Disorder as Revealed through Weekly Text Messaging: A Feasibility Study," *Bipolar Disorders* 12, no. 3 (2010): 327–34, doi:10.1111/j.1399-5618.2010.00807.x.; Z. Kupper and H. Hoffmann, "Course Patterns of Psychosocial Functioning in Schizophrenia Patients Attending a Vocational Rehabilitation Program," *Schizophrenia Bulletin* 26, no. 3 (2000): 681–98.

6. Rui Chen, George I. Mias, Jennifer Li-Pook-Than, Lihua Jiang, Hugo Y. K. Lam, Rong Chen, Elana Miriami, Konrad J. Karczewski, Manoj Hariharan, Frederick E. Dewey, et al., "Personal Omics Profiling Reveals Dynamic Molecular and Medical Phenotypes," *Cell* 148, no. 6 (2012): 1293–307, doi:10.1016/j.cell.2012.02.009.

7. Allen B. Weisse, "Self-Experimentation and Its Role in Medical Research," *Texas Heart Institute Journal* 39, no. 1 (2012): 51–54.

8. Timothy O. Laumann, Evan M. Gordon, Babatunde Adeyemo, Abraham Z. Snyder, Sung Jun Joo, Mei-Yen Chen, Adrian W. Gilmore, Kathleen B. McDermott, Steven M. Nelson, Nico U. F. Dosenbach, et al., "Functional System and Areal Organization of a Highly Sampled Individual Human Brain," *Neuron* 87, no. 3 (2015): 657–70, doi:10.1016/j.neuron.2015.06.037.

9. Rodrigo M. Braga and Randy L. Buckner, "Parallel Interdigitated Distributed Networks within the Individual Estimated by Intrinsic Functional Connectivity," *Neuron* 95, no. 2 (2017): 457–71.e5, doi: 10.1016/j.neuron.2017.06.038.

Chapter 6

1. *Graham v. Florida*, 560 US 1, 17 (2010).
2. William Shakespeare, "The Winter's Tale," act 3, scene 3, in *The Complete works of William Shakespeare* (MIT, Internet, n.d.), http://shakespeare.mit.edu/winters_tale/winters_tale.3.3.html.
3. Adriana Galvan, Todd A. Hare, Cindy E. Parra, Jackie Penn, Henning Voss, Gary Glover, and B. J. Casey, "Earlier Development of the Accumbens Relative to Orbitofrontal Cortex Might Underlie Risk-Taking Behavior in Adolescents," *Journal of Neuroscience* 26, no. 25 (2006): 6885–92, doi:10.1523/JNEUROSCI.1062-06.2006.
4. Jessica R. Cohen, Robert F. Asarnow, Fred W. Sabb, Robert M. Bilder, Susan Y. Bookheimer, Barbara J. Knowlton, and Russell A. Poldrack, "A Unique Adolescent Response to Reward Prediction Errors," *Nature Neuroscience* 13, no. 6 (2010): 669–71, doi:10.1038/nn.2558.
5. Jeffrey M. Burns and Russell H. Swerdlow, "Right Orbitofrontal Tumor with Pedophilia Symptom and Constructional Apraxia Sign," *Archives of Neurology* 60, no. 3 (2003): 437–40.
6. *Frye v. United States*, 293 F, 1013, 1014 (D.C. Cir. 1923).
7. *Daubert v. Merrell Dow Pharmaceuticals, Inc.*, 509 US 579, 580 (1993).
8. Committee to Review the Scientific Evidence on the Polygraph; Board on Behavioral, Cognitive, and Sensory Sciences and Committee on National Statistics; Division of Behavioral and Social Sciences and Education; National Research Council of the National Academies, *The Polygraph and Lie Detection* (Washington, DC: National Academies Press, 2003).
9. D. D. Langleben, L. Schroeder, J. A. Maldjian, R. C. Gur, S. McDonald, J. D. Ragland, C. P. O'Brien, and A. R. Childress, "Brain Activity During Simulated Deception: An Event-Related Functional Magnetic Resonance Study," *Neuroimage* 15, no. 3 (2002): 727–32.
10. C. Davatzikos, K. Ruparel, Y. Fan, D. G. Shen, M. Acharyya, J. W. Loughead, R. C. Gur, and D. D. Langleben, "Classifying Spatial Patterns of Brain Activity with Machine Learning Methods: Application to Lie Detection," *Neuroimage* 28, no. 3 (2005): 663–68.
11. F. A. Kozel, K. A. Johnson, Q. Mu, E. L. Grenesko, S. J. Laken, and M. S. George, "Detecting Deception Using Functional Magnetic Resonance Imaging," *Biological Psychiatry* 58, no. 8 (2005): 605–13.
12. *United States v. Semrau*, No. 11-5396 (6th Cir. 2012), http://www.opn.ca6.uscourts.gov/opinions.pdf/12a0312p-06.pdf.
13. Robert Huizenga, "Dr. Oz and Dr. H: Behind the Murder Case—The Gary Smith Story," *YouTube*, March 18, 2016, https://www.youtube.com/watch?v=-JB4jRV_38E. Quote begins at time point 0:53.
14. Robert Huizenga, personal communication, June 18, 2017.
15. "The Beard," episode of *Seinfeld*, broadcast February 9, 1995 (NBC).
16. G. Ganis, J. P. Rosenfeld, J. Meixner, R. A. Kievit, and H. E. Schendan, "Lying in the Scanner: Covert Countermeasures Disrupt Deception Detection by Functional Magnetic Resonance Imaging," *Neuroimage* 55, no. 1 (2011): 312–19.
17. Seena Fazel, Zheng Chang, Thomas Fanshawe, Niklas Långström, Paul Lichtenstein, Henrik Larsson, and Susan Mallett, "Prediction of Violent Reoffending on Release from Prison: Derivation and External Validation of a Scalable Tool," *Lancet Psychiatry* 3, no. 6 (2016): 535–43, doi:10.1016/S2215-0366(16)00103-6.

18. E. Aharoni, G. M. Vincent, C. L. Harenski, V. D. Calhoun, W. Sinnott-Armstrong, M. S. Gazzaniga, and K. A. Kiehl, "Neuroprediction of Future Rearrest," *Proceedings of the National Academy of Sciences of the USA* 110, no. 15 (2013): 6223–28.

19. Russell A. Poldrack, "How Well Can We Predict Future Criminal Acts from fMRI Data?" *RussPoldrack.org* (blog), April 6, 2013, http://www.russpoldrack.org/2013/04/how-well-can-we-predict-future-criminal.html.

20. Derek W. Braverman, Samuel N. Doernberg, Carlisle P. Runge, and Dana S. Howard "OxRec Model for Assessing Risk of Recidivism: Ethics," *Lancet Psychiatry* 3, no. 9 (2016): 808–9, doi:10.1016/S2215-0366(16)30175-4.

21. Vinayak K. Prasad and Adam S. Cifu, *Ending Medical Reversal: Improving Outcomes, Saving Lives* (Baltimore, MD: Johns Hopkins University Press).

22. John P. A. Ioannidis, "Why Most Published Research Findings Are False," *PLoS Medicine* 2, no. 8 (2005): e124, doi:10.1371/journal.pmed.0020124.

23. J. P. Simmons, L. D. Nelson, and U. Simonsohn, "False-Positive Psychology: Undisclosed Flexibility in Data Collection and Analysis Allows Presenting Anything as Significant," *Psychological Science* 22, no. 11 (2011): 1359–66.

24. Ed Yong, "Replication Studies: Bad Copy," *Nature* 485, no. 7398 (2012): 298–300, doi:10.1038/485298a.

25. Open Science Collaboration, "PSYCHOLOGY: Estimating the Reproducibility of Psychological Science," *Science* 349, no. 6251 (2015): aac4716, doi: 10.1126/science .aac4716.

26. Ed Yong, "How Reliable Are Psychology Studies?" *Atlantic*, August 27, 2015, https://www.theatlantic.com/science/archive/2015/08/psychology-studies-reliability -reproducability-nosek/402466/.

27. Katherine S. Button, John P. A. Ioannidis, Claire Mokrysz, Brian A. Nosek, Jonathan Flint, Emma S. J. Robinson, and Marcus R. Munafò, "Power Failure: Why Small Sample Size Undermines the Reliability of Neuroscience," *Nature Reviews Neuroscience* 14, no. 5 (2013): 365–76, doi: 10.1038/nrn3475.

28. Russell A. Poldrack, Chris I. Baker, Joke Durnez, Krzysztof J. Gorgolewski, Paul M. Matthews, Marcus R. Munafò, Thomas E. Nichols, Jean-Baptiste Poline, Edward Vul, and Tal Yarkoni, "Scanning the Horizon: Towards Transparent and Repro- ducible Neuroimaging Research," *Nature Reviews Neuroscience* 18 (2017): 115–26, doi:10.1038/nrn.2016.167.

Chapter 7

1. The Notorious B.I.G., featuring Puff Daddy and Mase, "Mo Money Mo Problems," *Life After Death* [CD] (Bad Boy/Arista, 1996).

2. Michael Lewis, *The Undoing Project: A Friendship That Changed Our Minds* (New York: W. W. Norton, 2017).

3. From an unpublished patient report kindly communicated by Bruce Miller and Robin Ketelle of the University of California, San Francisco.

4. Sabrina M. Tom, Craig R. Fox, Christopher Trepel, and Russell A. Poldrack, "The Neural Basis of Loss Aversion in Decision-Making under Risk," *Science* 315, no. 5811 (2007): 515–18, doi:10.1126/science.1134239.

5. Nicola Canessa, Chiara Crespi, Matteo Motterlini, Gabriel Baud-Bovy, Gabriele Chierchia, Giuseppe Pantaleo, Marco Tettamanti, and Stefano F. Cappa, "The Functional and Structural Neural Basis of Individual Differences in Loss Aversion," *Journal of Neuroscience* 33, no. 36 (2013): 14307–17, doi:10.1523/JNEUROSCI.0497- 13.2013.

6. Emily A. Ferenczi, Kelly A. Zalocusky, Conor Liston, Logan Grosenick, Melissa R. Warden, Debha Amatya, Kiefer Katovich, Hershel Mehta, Brian Patenaude, Charu Ramakrishnan, et al., "Prefrontal Cortical Regulation of Brainwide Circuit Dynamics and Reward-Related Behavior," *Science* 351, no. 6268 (2016): aac9698, doi:10.1126/science.aac9698.

7. John Maynard Keynes, *The General Theory of Employment, Interest and Money* (New York: Harcourt Brace, 1936), 161.

8. Daniel Kahneman, *Thinking, Fast and Slow* (New York: Farrar, Straus and Giroux, 2013).

9. William James, *The Principles of Psychology* (New York: H. Holt 1890), 122.

10. D. Shohamy, C. E. Myers, S. Grossman, J. Sage, M. A. Gluck, and R. A. Poldrack, "Cortico-striatal Contributions to Feedback-Based Learning: Converging Data from Neuroimaging and Neuropsychology," *Brain* 127, no. 4 (2004): 851–59, doi:10.1093/brain/awh100.

11. R. A. Poldrack, J. Clark, E. J. Paré-Blagoev, D. Shohamy, J. Creso Moyano, C. Myers, and M. A. Gluck, "Interactive Memory Systems in the Human Brain," *Nature* 414, no. 6863 (2001): 546–50, doi: 10.1038/35107080.

12. Kevin J. Klos, James H. Bower, Keith A. Josephs, Joseph Y. Matsumoto, and J. Eric Ahlskog, "Pathological Hypersexuality Predominantly Linked to Adjuvant Dopamine Agonist Therapy in Parkinson's Disease and Multiple System Atrophy," *Parkinsonism and Related Disorders* 11, no. 6 (2005): 381–86, doi:10.1016/j.parkreldis.2005.06.005.

13. Samuel M. McClure, David Laibson, George Loewenstein, and Jonathan D. Cohen, "Separate Neural Systems Value Immediate and Delayed Monetary Rewards," *Science* 306, no. 5695 (2004): 506, doi:10.1126/science.1100907.

14. Joseph W. Kable and Paul W. Glimcher, "The Neural Correlates of Subjective Value during Intertemporal Choice," *Nature Neuroscience* 10, no. 12 (2007): 1625–33, doi:10.1038/nn2007.

15. W. Mischel, Y. Shoda, and M. I. Rodriguez, "Delay of Gratification in Children," *Science* 244, no. 4907 (1989): 933–38.

16. Adam R. Aron, "The Neural Basis of Inhibition in Cognitive Control," *Neuroscientist* 13, no. 3 (2007): 214–28, doi:10.1177/1073858407299288.

17. Samuel M. McClure, Jian Li, Damon Tomlin, Kim S. Cypert, Latané M. Montague, and P. Read Montague, "Neural Correlates of Behavioral Preference for Culturally Familiar Drinks," *Neuron* 44, no. 2 (2004): 379–87, doi:10.1016/j.neuron.2004 .09.019.

18. Marco Iacoboni, "Who Really Won the Super Bowl?" *Edge*, February 2, 2006, https://www.edge.org/conversation/marco_iacoboni-who-really-won-the-super-bowl.

19. Martin Lindstrom, "You Love Your iPhone. Literally," *New York Times*, September 30, 2011, http://www.nytimes.com/2011/10/01/opinion/you-love-your-iphone-literally .html.

20. Russell Poldrack, "The iPhone and the Brain," *New York Times*, "Opinion," October 4, 2011, http://www.nytimes.com/2011/10/05/opinion/the-iphone-and-the-brain.html.

21. Vinod Venkatraman, Angelika Dimoka, Paul A. Pavlou, Khoi Vo, William Hampton, Bryan Bollinger, Hal E. Hershfield, Masakazu Ishihara, and Russell S. Winer, "Predicting Advertising Success beyond Traditional Measures: New Insights from Neurophysiological Methods and Market Response Modeling," *Journal of Marketing Research* 52, no. 4 (2015): 436–52.

22. Emily B. Falk, Elliot T. Berkman, and Matthew D. Lieberman, "From Neural Responses to Population Behavior: Neural Focus Group Predicts Population-Level Media Effects," *Psychological Science* 23, no. 5 (2012): 439–45, doi:10.1177/0956797611434964.

23. Alexander Genevsky and Brian Knutson, "Neural Affective Mechanisms Predict Market-Level Microlending," *Psychological Science* 26, no. 9 (2015): 1411–22, doi:10.1177/0956797615588467.

Chapter 8

1. Thomas H. McGlashan, "Psychosis as a Disorder of Reduced Cathectic Capacity: Freud's Analysis of the Schreber Case Revisited," *Schizophrenia Bulletin* 35, no. 3 (2009): 478, doi:10.1093/schbul/sbp019.

2. Wendy Johnson, Lars Penke, and Frank M. Spinath, "Heritability in the Era of Molecular Genetics: Some Thoughts for Understanding Genetic Influences on Behavioural Traits," *European Journal of Personality* 25, no. 4 (2011): 254–66.

3. Aswin Sekar, Allison R. Bialas, Heather de Rivera, Avery Davis, Timothy R. Hammond, Nolan Kamitaki, Katherine Tooley, Jessy Presumey, Matthew Baum, Vanessa Van Doren, et al., "Schizophrenia Risk from Complex Variation of Complement Component 4," *Nature* 530, no. 7589 (2016): 177–83, doi:10.1038/nature16549.

4. Madeleine Goodkind, Simon B. Eickhoff, Desmond J. Oathes, Ying Jiang, Andrew Chang, Laura B. Jones-Hagata, Brissa N. Ortega, Yevgeniya V. Zaiko, Erika L. Roach, Mayuresh S. Korgaonkar, et al., "Identification of a Common Neurobiological Substrate for Mental Illness," *JAMA Psychiatry* 72, no. 4 (2015): 305–15, doi:10.1001/jamapsychiatry.2014.2206.

5. E. Sprooten, A. Rasgon, M. Goodman, A. Carlin, E. Leibu, W. H. Lee, and S. Frangou, "Addressing Reverse Inference in Psychiatric Neuroimaging: Meta-analyses of Task-Related Brain Activation in Common Mental Disorders," *Human Brain Mapping* 38, no. 4 (2017): 1846–64.

6. Jong H. Yoon, Andrew J. Westphal, Michael J. Minzenberg, Tara Niendam, J. Daniel Ragland, Tyler Lesh, Marjorie Solomon, and Cameron S. Carter, "Task-Evoked Substantia Nigra Hyperactivity Associated with Prefrontal Hypofunction, Prefronton-igral Disconnectivity and Nigrostriatal Connectivity Predicting Psychosis Severity in Medication Naïve First Episode Schizophrenia," *Schizophrenia Research* 159, no. 2–3 (2014): 521–26, doi: 10.1016/j.schres.2014.09.022.

7. American Psychiatric Association, *Diagnostic and Statistical Manual of Mental Disorders*, 5th ed. (Arlington, VA: American Psychiatric Publishing, 2013), 300.01 (F41.0).

8. Thomas Insel, "Transforming Diagnosis," *National Institute of Mental Health, Posts by Former NIMH Director Thomas Insel* [blog], April 29, 2013, https://www.nimh.nih.gov/about/directors/thomas-insel/blog/2013/transforming-diagnosis.shtml.

9. Jennifer S. Stevens, Ye Ji Kim, Isaac R. Galatzer-Levy, Renuka Reddy, Timothy D. Ely, Charles B. Nemeroff, Lauren A. Hudak, Tanja Jovanovic, Barbara O. Rothbaum, and Kerry J. Ressler, "Amygdala Reactivity and Anterior Cingulate Habituation Predict Posttraumatic Stress Disorder Symptom Maintenance after Acute Civilian Trauma," *Biological Psychiatry* 81, no. 12 (2017): 1023–29.

10. Michael J. Frank, Lauren C. Seeberger, and Randall C. O'Reilly, "By Carrot or by Stick: Cognitive Reinforcement Learning in Parkinsonism," *Science* 306, no. 5703 (2004): 1940–43, doi:10.1126/science.1102941.

11. Tiago V. Maia and Michael J. Frank, "An Integrative Perspective on the Role of Dopamine in Schizophrenia," *Biological Psychiatry* 81, no. 1 (2017): 52–66, doi:10.1016/j.biopsych.2016.05.021.

12. NIDA, "Drug Abuse and Addiction," in *Drugs, Brains, and Behavior: The Science of Addiction* (NIDA, July 1, 2014), https://www.drugabuse.gov/publications/drugs-brains -behavior-science-addiction/drug-abuse-addiction.

13. L. N. Robins, "The Sixth Thomas James Okey Memorial Lecture. Vietnam Veterans' Rapid Recovery from Heroin Addiction: A Fluke or Normal Expectation?," *Addiction* 88, no. 8 (1993): 1041–54.

14. Shepard Siegel, "Pavlovian Conditioning and Drug Overdose: When Tolerance Fails," *Addiction Research and Theory* 9, no. 5 (2001): 503–513, doi:10.3109/16066350109141767.

15. Nicolas Rüsch, Matthias C. Angermeyer, and Patrick W. Corrigan, "Mental Illness Stigma: Concepts, Consequences, and Initiatives to Reduce Stigma," *European Psychiatry* 20, no. 8 (2005): 529–39, doi:10.1016/j.eurpsy.2005.04.004.

16. Georg Schomerus, Christian Schwahn, Anita Holzinger, Patrick William Corrigan, Hans Jörgen Grabe, Mauro Giovanni Carta, and Matthias Claus Angermeyer, "Evolution of Public Attitudes about Mental Illness: A Systematic Review and Meta-analysis," *Acta Psychiatrica Scandinavica* 125, no. 6 (2012): 440–52, doi:10.1111/j.1600-0447.2012.01826.x.

17. Nick Haslam and Erlend P. Kvaale, "Biogenetic Explanations of Mental Disorder," *Current Directions in Psychological Science* 24, no. 5 (2015): 399–404, doi:10.1177/0963721415588082.

18. Carla Meurk, Kylie Morphett, Adrian Carter, Megan Weier, Jayne Lucke, and Wayne Hall, "Scepticism and Hope in a Complex Predicament: People with Addictions Deliberate about Neuroscience," *International Journal on Drug Policy* 32 (2016): 34–43, doi:10.1016/j.drugpo.2016.03.004. Quotes 38–39.

Chapter 9

1. Essa Yacoub, Noam Harel, and Kâmil Ugurbil, "High-Field fMRI Unveils Orientation Columns in Humans," *Proceedings of the National Academy of Sciences of the USA* 105, no. 30 (2008): 10607–12, doi:10.1073/pnas.0804110105.

2. Dan L. Longo and Jeffrey M. Drazen, "More on Data Sharing," *New England Journal of Medicine* 374, no. 19 (2016): 1896–97, doi:10.1056/NEJMc1602586.

3. Bharat B. Biswal, Maarten Mennes, Xi Nian Zuo, Suril Gohel, Clare Kelly, Steve M. Smith, Christian F. Beckmann, Jonathan S. Adelstein, Randy L. Buckner, Stan Colcombe, et al., "Toward Discovery Science of Human Brain Function," *Proceedings of the National Academy of Sciences of the USA* 107, no. 10 (2010): 4734–39, doi:10.1073/pnas.0911855107.

4. Karla L. Miller, Fidel Alfaro-Almagro, Neal K. Bangerter, David L. Thomas, Essa Yacoub, Junqian Xu, Andreas J. Bartsch, Saad Jbabdi, Stamatios N. Sotiropoulos, Jesper L. R. Andersson, et al., "Multimodal Population Brain Imaging in the UK Biobank Prospective Epidemiological Study," *Nature Neuroscience* 19, no. 11 (2016): 1523–36, doi:10.1038/nn.4393.

5. See http://www.openfmri.org.
6. The only information that I have not released is the contents of the diaries that I kept after each scan, since they contain potentially sensitive information about others.
7. Richard F. Betzel, Theodore D. Satterthwaite, Joshua I. Gold, and Danielle S. Bassett, "Positive Affect, Surprise, and Fatigue Are Correlates of Network Flexibility," *Scientific Reports* 7, no. 1 (2017): 520, doi:10.1038/s41598-017-00425-z.

INDEX